A Critique of the Arguments for Scientific Realism

Dr Phil Rees

Copyright © 2012 by Phil Rees

All rights reserved. This book, or parts thereof, may not be reproduced in any form without permission.

A catalogue record for this book is available from the British Library

ISBN: 978-1-907962-51-6

Published by Cranmore Publications
Reading, England

www.cranmorepublications.co.uk

Abstract

This book aims to show that the arguments offered in support of the most prevalent versions of scientific realism are unconvincing and offer no reason why an anti-realist position should not rationally be held.

Several formulations of realism are examined and reduced to two. First, Convergent Ontological Scientific Realism (COSR) – the claim that most current successful theories are approximately true, and successive theories converge upon the truth. Second, Axiological Scientific Realism (ASR) – the claim that while many current theories may not even be approximately true, nevertheless, science *aims* for the truth.

Axiological claims of all kinds are questioned and shown to be unhelpful. It is then argued that ASR theorists face difficulties, and need to also affirm COSR. Consequently, refuting COSR refutes ASR, and that is what the remainder of the book attempts to do.

The concept of approximate truth is critiqued and shown to be incapable of doing the relatively precise job scientific realism requires of it.

The *Pessimistic Induction* argument is presented in a revised form, and an extensive selection of realist criticisms are examined and rebutted. The claim against realism remains intact – that the historical record refutes its proposed link between empirical success and approximate truth.

The classical *Underdetermination* argument against realism is criticised. However, the *Unconceived Alternatives* argument is shown to undermine scientific realism by an induction on the historical record.

The *No Miracles Argument* for scientific realism is examined in detail and shown to be question-begging, and unconvincing on many other counts.

A non-realist alternative explanation of the success of science is offered which explains more than the putative realist explanation would have done.

The conclusion is that the case for scientific realism has not been made, leaving the way open for non-realist ways of understanding the activities of the sciences.

Contents

ABSTRACT 3
CONTENTS 5
ACKNOWLEDGEMENTS 7
§1 INTRODUCTION: THE REALIST TURN 9
§2 WHAT IS SCIENTIFIC REALISM? 25
§3 AXIOLOGICAL SCIENTIFIC REALISM (ASR) 54
§4 VERISIMILITUDE, APPROXIMATE TRUTH, AND
    TRUTH-LIKENESS 94
§5 THE PESSIMISTIC INDUCTION ARGUMENT 116
§6 DEFENDING THE PESSIMISTIC INDUCTION 138
§7 UNDERDETERMINATION AND UNCONCEIVED
    ALTERNATIVES 207
§8 THE NO MIRACLES ARGUMENT FOR SCIENTIFIC
    REALISM 244
§9 EXPLAINING THE SUCCESS OF SCIENCE 281
§10 CONCLUSIONS 313
GLOSSARY OF TERMS AND ABBREVIATIONS 321
APPENDIX 1 : VARIETIES OF SCIENTIFIC REALISM 325
APPENDIX 2 : THE BASE RATE FALLACY ARGUMENT 336
REFERENCES 342

ACKNOWLEDGEMENTS

I wish to acknowledge the help I have received throughout my work on this book from John Preston whose persistent encouragement and constructive criticism has played a major role in my bringing this work to fruition. I have greatly benefitted from discussions with Mike Arnautov, Alice Drewery, Peter Townsend, and from email correspondence with Bas van Fraassen, Timothy Lyons, Alan Musgrave, Stathis Psillos, Kyle Stanford. Jon Beale read through a draft of the entire essay and I am grateful for the many helpful suggestions he made. I am also grateful for comments received at Reading University graduate seminars, particularly from Alex Gregory, Jonathan Powell, Chris Pulman, Jussi Suikkanen. Finally, I must record my gratitude to, and appreciation of, the work of van Fraassen; reading his *The Empirical Stance* (2002) initiated a thorough change in my philosophical outlook for which I am immensely grateful.

§1 Introduction: The Realist Turn

§1.1 Introductory Remarks

This essay is concerned with the debate between realism and anti-realism in the philosophy of science and could be regarded as the first of a two-part project, the second part of which would set out and defend a specific anti-realist position. However, the aim here is to show that the arguments offered in support of the most prevalent versions of scientific realism are unconvincing and offer no reason why an anti-realist position should not rationally be held.

In this chapter I will introduce, in broad terms, the two positions, and discuss the historical background to the debate. In the next chapter I evaluate in detail various actual formulations of scientific realism, and describe what I shall take the realist position(s) to be.

Here are some typical questions within the philosophy of science: What are the aims of science? What does science achieve? Do scientific theories describe the world? Should they? If science delivers knowledge, what is it knowledge of? In addition to their importance, how one answers these questions will indicate one's position in the realism/anti-realism debate, and thus it is not difficult to see how that debate comes to occupy such a central position within the discipline.

The replies of many realists to the first of these questions will include the claim that science aims to discover the truth concerning a 'hidden reality' that explains the phenomena we observe. Other realists will claim that in many cases science has already achieved this aim. Common to both groups will be the assumption of that explanatory hidden reality, and then the knowledge delivered by science will be knowledge of that hidden reality, and scientific theories will be said to describe the world or aim to do so. For the realist, science is concerned with gaining ever deeper explanations of the world's phenomena, with the assumption that much of the world is hidden to us, in the sense of not being directly available to experience. The realist holds that the behaviour of the objects and phenomena of direct experience can be explained only by assuming the literal existence of entities, properties, and processes that are not accessible in experience, and that these are what the theoretical terms of scientific theories refer to.

The anti-realist will deny this and may hold that the business of science is the succinct codification of the world's directly perceived phenomena in a form that, together with an assumption of continuity and uniformity, enables accurate predictions of those phenomena to be made.[1] On this account, the knowledge delivered by science will be knowledge of what phenomena actually occur and what phenomena may be expected to occur in other places and times. The realist thinks of science as a voyage of discovery, but although the anti-realist *may* concede the possibility of some such discovery, she will wish to view science as *primarily* concerned to accurately describe the world's phenomena, thus enabling successful interventions in the world.

It is often assumed that most modern scientists side with the realist, believing that their theories reveal some hidden portion of the world; that science revises our picture of how 'reality' is constituted.[2] Regardless of what scientists may

---

[1] Heinrich Hertz would certainly agree: 'The most direct ... problem which our conscious knowledge of nature should enable us to solve is the anticipation of future events, so that we may arrange our present affairs in accordance with such anticipation.' (1900: p1).

[2] The writings of some scientific realists seem to take this for granted, for example this by Richard Boyd: 'Scientific realism is thus the common sense ... conception that we are justified in accepting the most secure findings of

§1 Introduction: The Realist Turn

believe, philosophers of science are divided over these questions and their arguments largely define the scientific realism/anti-realism debate. Realists hold that theories are intended to accurately describe the world, including that portion which lies beyond experience; and that they progressively approximate to achieving this. By contrast, as I have said, anti-realists hold that science is concerned to describe the behaviour of the world in a manner that enables accurate predictions to be made, thus enabling pragmatically useful interventions to occur. It is not that anti-realists wish to contradict those realist scientists or tell them they are mistaken, but rather they propose that it is possible to take a different stance towards science – an alternative way of viewing and understanding the scientific enterprise – a way which, they claim, avoids what they see as needless and risky commitments to entities, properties and processes that go beyond experience.

While connections could be established, the topic of scientific realism/anti-realism with which I am concerned bears little relation to what I shall call 'global realism/anti-realism'. The terms realism and anti-realism can be (and are) applied to a wide variety of philosophical situations – for example, ethics, modality, mathematics, consciousness. In its global form,

--- footnote continuation

scientists "at face value"' (2002: preface). This is dubious as it begs the question as to what scientists think are the 'face value' findings of their work. As we will see below, nineteenth-century scientists were distinctly anti-realist, and Newton thought his work revealed divine agency at work in the world. The right way to understand the words of scientists is not by projecting naïve realism onto them. This remark of physicist Steven Weinberg illustrates the ambiguity present in the attitudes of many scientists as to the 'face value' of their findings: 'When we say that a thing is real we are simply expressing a sort of respect. We mean that the thing must be taken seriously because we cannot learn about it without making an effort that goes beyond our imagination. ... But I have to admit that my willingness to grant the title of "real" is a little like Lloyd George's willingness to grant titles of nobility; it is a measure of how little difference I think the title makes' (1993: pp35-6).

realism is opposed to idealism, which has long been out of favour in contemporary philosophy. However, in more recent times the debate concerning global realism/anti-realism has largely centred around Michael Dummett's transformation of the topic from a question of metaphysics to one within the philosophy of language. Dummett proposes that global realism/anti-realism amounts to a semantic dispute about the kind of truth conditions that should apply to the sentences used. According to his semantic realism, the truth conditions are verification-transcendent in that they obtain despite the fact that we may be unable to decide whether they obtain. On the other hand, for Dummett's semantic anti-realist the truth conditions are of a verificationist kind, and truth cannot outrun all possible human knowledge. Global realism/anti-realism will not concern me here, though I will refer to it in the historical discussion below. It is sometimes referred to as 'metaphysical realism/anti-realism', but I will not use that phrase.

I will examine in considerable detail the actual claims of scientific realism and the arguments put forward in its favour, and I will show that these arguments fail, while also showing that, contrary to the claims of realists, various arguments in favour of anti-realism have not been refuted, and represent a very serious challenge to realism. The conclusion will be that scientific realists fail to make their case and offer nothing to show that it is not perfectly rational to adopt a more sceptical position regarding the theoretical commitments of current science.

In this chapter I will give a brief outline of the historical context of the realism/anti-realism debate, together with an equally brief outline of my own position within that debate.

§1.2 The Historical Background: The Realist Turn

Let me first place the debate in its historical context. Before Kant, science and substantive metaphysics were inextricably linked.[3] We may recall Descartes's metaphor of philosophy as a tree whose roots are metaphysics, whose trunk is physics, and whose branches are the special sciences, but Kant undermined the legitimacy of a metaphysics that is completely removed from experience. After Kant, in some schools, metaphysics was no longer taught at all, and the nineteenth-century saw science increasingly free of metaphysics, with discussions of Natural Kinds, Universals, Laws of Nature as necessitating, all losing their place in the scientific conception of the world. This trend reached a peak with Pierre Duhem and Ernst Mach, the former beginning the quest for clear lines of demarcation between the two disciplines, while the latter began the process that culminated in the Vienna Circle's assertion of the meaninglessness of all metaphysical statements. Russell even dismissed causation as 'a relic of a bygone age' (1912b: p193).[4]

By the late nineteenth-century the philosophical thinking of scientists about science was dominated by various forms of what would now be described as anti-realism. Many of the greatest scientists were also philosophers of science – one thinks here of Maxwell, Boltzmann, Hertz, Helmholtz, Kekulé, Mach. Here are three quotes that indicate the viewpoints of these people, viewpoints we would today take to be non-realist:

---

[3] I use the word 'substantive' in the sense illustrated here: 'Some metaphysical projects have no ambitions beyond conceptual analysis, but others make substantive claims about the nature of reality'.
[4] Mach (1885) would have agreed, referring to it as a 'fetish'.

§1 Introduction: The Realist Turn

Kekulé (1867):

> The question whether atoms exist or not has little significance in a chemical point of view : its discussion belongs to metaphysics. In chemistry we have to decide whether the assumption of atoms is a hypothesis adapted to the explanation of chemical phenomena. ... I have no hesitation in saying that, from a philosophical point of view, I do not believe in the actual existence of atoms ...

Boltzmann (1905: p252):

> What the real cause for the fact that the world of appearance runs its course in just this way may be; what may be hidden behind the world of appearance, propelling it, as it were - such investigations we do not consider to be of the task of natural science.[5]

Hertz on Maxwell's theory (1893: p28):

> If we wish to lend more colour to our theory, there is nothing to prevent us from supplementing all this and aiding our powers of imagination by concrete representations of the various conceptions as to the nature of electric polarisation, the electric current, etc. But scientific accuracy requires of us that we should in no wise confuse the simple and homely figure, as it is presented to us by nature, with the gay garment which we use to clothe it. Of our own free will we can make no change whatever in the form of the one, but the cut and colour of the other we can choose as we please.

---

[5] But note that the way this is stated suggests that Boltzmann thinks there is something 'hidden behind the world of appearance, propelling it, as it were', but that it is not the job of science to investigate this. This seems a rather Duhemian view, and not characteristic of Boltzmann's usually more verificationist stance.

§1 Introduction: The Realist Turn

Mainstream philosophy itself was almost entirely idealist, both on the continent and in the UK and America, so there was little sign of what I have called *global realism*. However, among scientists themselves this began to break down with Max Planck's turn of the century attacks on Ernst Mach's phenomenalism, culminating in his 1908 address to the Faculty of Natural Sciences of the University of Leiden which was a polemic against Mach and anti-realism. This current of criticism of anti-realism had accelerated after Einstein's 1905 paper on Brownian motion which was taken as supportive of a realist view of atoms.[6] However, within a couple of decades, this realist trend within science itself was checked by problems associated with the interpretation of quantum theory. A realist interpretation was rejected by the scientific community in favour of the instrumentalist programme promoted by Bohr and Heisenberg.

By this time mainstream philosophy had turned against Idealism, though Russell's professed realism was equivocal, with the shadows of Mach and Kant still showing non-realist influences.[7] Moreover, philosophy itself continued to resist global realism when the Vienna Circle's positivism led it to declare the whole issue of realism a metaphysical pseudo-question. Additionally, Wittgenstein's *Tractatus*, which so heavily influenced the Vienna Circle, shows many traces of the philosophy that suggested scientific theories are best conceived of as models or *pictures*, rather than as truth

---

[6] After Einstein's 1905 publication of his theoretical explanation of Brownian motion in terms of molecules, Jean Perrin did the experimental work to test Einstein's predictions, publishing the results in his 1909 paper 'Brownian Movement and Molecular Reality' which introduced the term "Avogadro's Number". Perrin's later book *Les Atomes* effected a decisive realist swing concerning atoms and molecules (see Nye, 1972).

[7] As an example of the Kantian influence, consider 'we can know nothing of what [physical space] is like in itself,...' (1912a: p15) and 'the physical objects themselves remain unknown in their intrinsic nature ...' (*ibid:* p17).

evaluable statements about reality.[8] Thus for a while empiricist and instrumentalist thought continued to eclipse realist thought in both philosophy and science.

It is a reasonable generalisation that in philosophy global realism and substantive metaphysics often go together, as do anti-realism and empiricism, with its rejection of such metaphysics. Realism seems to demand at least some such metaphysics due to its two claims that 'reality' outstrips appearance, and that we can have knowledge of that reality. The former raises the issue of an evidence-transcendent truth, and the latter the problem of how to account for knowledge of something that goes beyond the appearances of which we do have knowledge. Clearly explanations of these mysteries cannot be given in terms of the appearances themselves, so no *a posteriori* empirical explanation can be had, and that leaves only *a priori* metaphysical theory, to which empiricism is opposed.

In discussing his opposition to such metaphysics, Bas van Fraassen (2002: pp36-7) depicts the history of philosophy as a series of empiricist attacks upon substantive metaphysics, from the medieval William of Occam's nominalist rejection of universals, to Hume's sceptical empiricism and his advocating the burning of books of metaphysics, and finally the extreme verificationism of the Vienna Circle, rendering literally meaningless all metaphysical utterances. Indeed, as van Fraassen also observes, this way of carving philosophy at the joints might even lead to casting Aristotle as an empiricist reaction against Plato's metaphysical extremes.

Following the disintegration of the Vienna Circle in the late 1930s, and the war interregnum, anglophone philosophy

---

[8] The influence of Hertz on Wittgenstein's 'Picture Theory' is widely acknowledged.

became preoccupied almost entirely with problems of language. Nevertheless, many of the Vienna Circle membership re-grouped in America and founded the discipline known as 'philosophy of science'. One thinks here of Rudolf Carnap, Hans Reichenbach, Carl Hempel, Herbert Feigl, Phillip Frank, and others. Unsurprisingly, the discipline they founded was concerned mainly with epistemology, logic, methodology, and hardly at all with metaphysical questions. Logical positivism had developed two specific forms of scientific anti-realism. These were 'semantic instrumentalism', according to which theoretical terms should not be taken as referring to anything, and 'theoretical reductionism', according to which talk of theoretical entities can be reduced to talk of observable entities, properties, or processes.[9] However, by the 1960s the positivist star was waning rapidly.

The situation had begun to change during the late 1950s, and the realist turn was supported (perhaps even initiated) by Popper's attack on instrumentalism (1956) which he criticised as unable to account for his own falsificationist methodology. Also in 1956, Wilfrid Sellars published his famous attempted demolition of the empiricist 'myth of the given'.[10]

---

[9] I discuss the terms 'observable' and 'theoretical' in §2.2.

[10] See Sellars (1956). His essay contains the beginnings of many important scientific realist strands, for example: 'I am quite prepared to say that the commonsense world of physical objects in Space and Time is unreal ... that in the dimension of describing and explaining the world, science is the measure of all things, of what is that it is, and of what is not that it is not.' (1956: p173). Here we see the realist conception of science as guide to what there is, as opposed to van Fraassen's conception of science as guide to successful method. In addition, the first part of this quote grants reality to the unobservable, while denying the reality of the observable, a theme continued in his (1960). The quote thus illustrates how one kind of scientific realism amounts to an anti-realism concerning the world of everyday objects and phenomena, almost a direct inversion of the empiricist position; and also possibly an inversion of the terms realism/anti-realism, for the so-called scientific anti-realist is profoundly realist about that world of everyday objects and phenomena – for her there is

In addition, science and technology had become an increasingly ubiquitous feature of western culture and we can see a quite dramatic turn toward scientific realism with a concomitant increased interest in various metaphysical issues, both in general, and specifically within the philosophy of science itself. Various reasons have been suggested for this. Thomas Kuhn's *Structure of Scientific Revolutions* was published in 1962 and some viewed it as threatening some kind of postmodern relativism and even a loss of rationality. A turn to realism and metaphysics might then be seen as an antidote to relativism.[11] However, this could not be the entire reason since two other notable realist publications occurred at roughly the same time. First, Grover Maxwell published his 'The Ontological Status of Theoretical Entities' (1962) as a realist riposte to Carnap's earlier 'The Methodological Character of Theoretical Concepts' (1956). Secondly, J.J.C. Smart published his *Philosophy and Scientific Realism* (1963) with its raising of the 'No Miracles Argument' for scientific realism, a topic that I shall discuss in detail in §8. More recently, Saul Kripke's enormously influential *Naming and Necessity* (1972) raised the issue of metaphysical essentialism and necessity. This has flourished and there has been much work, and a growing literature, on scientific essentialism,[12] and it is now fair to say that substantive metaphysics, allied to a thoroughgoing realism, is very much in the ascendancy.

──────────────────────── footnote continuation
nothing more real and immediate. She, arguably, deserves the description 'realist' more than someone like Sellars who professes that title.

[11] George Reisch (1991) argues that Carnap admired Kuhn's work on revolutionary change, and that he expressed similar views. However, that doesn't negate the fact that the general reception of Kuhn's work was bad for logical positivism. Reisch himself says '*The Structure of Scientific Revolutions* rides atop a wave of reaction among many philosophers and historians of science to the logical empiricist program' (p264).

[12] For example in the work of Ellis (2001, 2002), Mumford (2004), Bird (2007). A more general defence of Aristotelian essentialism can be found in David Oderberg (2007).

There may also be deeper social reasons for this more recent turn to scientific realism. For example, the turn away from religious observance in the latter half of the twentieth-century in those countries where philosophy of science is practised may parallel an increased wish to see science as the new source of insight into the 'nature of reality', what lies behind it, and how it came to exist. Perhaps the religious world view giving way to the scientific world view must inevitably be accompanied by an increasing assumption that the latter needs a realist interpretation if it is to provide explanations that the former had been seen as providing. Humans seem to have a need to believe there is some unifying principle that can offer explanations in terms of hidden entities and mechanisms. This need was satisfied by the various deities who were the source of ultimate explanation, but now we live in an age where our faith in those sources of explanation has faded. It is therefore no surprise that science should increasingly come to be seen as the new source of ultimate explanation. The physicist's talk of a 'grand unifying theory' can sound very much like an ultimate explanation. Physicists also speak of understanding the fundamental nature of everything, and of the way the universe was created; of recreating, if only for nanoseconds, 'conditions as they were when the universe was created'. This also sounds like the language of ultimate explanation, and it is hardly surprising if such talk becomes the *lingua franca* of our culture. In these circumstances it is unsurprising if we find a turn to a thoroughgoing realist view of science.

Nevertheless, it would be a mistake to think that scientific anti-realism has been completely ousted and replaced with a realist hegemony; for it received a considerable boost with the publication of van Fraassen's *The Scientific Image* (1980). This proposed the explicitly anti-realist philosophy of *Constructive Empiricism* and represented a return to empiricism, but in a new form that was no longer reliant upon those older linguistic assumptions of the positivists. van

§1 Introduction: The Realist Turn

Fraassen himself said 'My own view is that empiricism is correct, but could not live in the linguistic form the positivists gave it' (p2). Since that book was published the debate between realism and anti-realism in the philosophy of science has continued for almost thirty years, and it is fair to say that scientific realism is by no means looking victorious. Indeed I will argue that it lacks credibility, at least in the way in which it has been presented. Some have suggested that the debate has reached a complete stalemate with each side offering arguments that beg the question against the other.[13] Be that as it may, I suggest that whilst the scientific realist research project has somewhat come to a standstill, the anti-realist project is currently attracting considerable interest. Much new and original work has been done, for example with Kyle Stanford's new approach to instrumentalism (2006), and also in the field of structuralism,[14] where both van Fraassen and Otávio Bueno continue to move forward with new proposals.[15]

---

[13] For example, Simon Blackburn (2002).

[14] Within the philosophy of science the word 'structuralism' refers broadly to the view that science can yield knowledge only of structure. Its more metaphysical version includes the additional claim that structure is all there *is* to know. It would be wrong to describe all versions of structuralism as anti-realist, but it would be reasonable to describe them as *non-realist* in the sense of the word *realist* considered in this essay. This point is developed more fully in §6.6.8 where I briefly discuss *Structural Realism*. The scientific structuralist concedes most of the problems with the kind of scientific realism that I discuss in this essay, but is determined to find something – structure – about which she can be realist. In terms of this essay I consider structuralists friends rather than foes! The literature on scientific structuralism is considerable, but examples of recent work are Ladyman (2002), French & Ladyman (forthcoming), da Costa & French (2003).

[15] Examples of recent explicitly anti-realist structuralist works are van Fraassen (2006a, 2008) who calls his position 'empiricist structuralism', and Bueno (1999, 2000) who uses the term 'structural empiricism'.

§1.3 The Position I Hold, but Do Not Defend Here

As I mentioned earlier, this essay will not set out and defend my own anti-realist position, as this is beyond the scope of this work. However, a brief but undefended indication of the main principles that guide my own thinking is appropriate.

As we have seen, scientific realism is today's orthodoxy and many current philosophers seem to assume that what science tells them is true or very close to the truth, and what is only 'very close to the truth' will soon have been developed into literal truth. I would advocate a scepticism that, in terms of our epistemic access to the world, sets the bar as high as reasonably possible. It is the attitude of the doubting Thomas who asks for verification and/or confirmation. This involves treating with extreme caution all claims as to the existence of entities/processes to which we can gain no independent confirmatory access.[16] Using an inference to best explanation (IBE) is fine in science where it results in a fallible hypothesis that can be subject to test. However, most uses of such inference in modern philosophy do not insist on such independent access, and thus simply *discover* necessity, self-identity, essences, and other such dubious items, leaving open the possibility that their discoveries are really inventions. I believe that this approach carries over into scientific realism where many philosophers are apt to show enormous trust in assuming that we are at, or close to, the truth. I advocate, instead, that we should always allow that the world may be, and probably is, very different from what our current scientific theories suggest. It is possible that future science will not include black holes, neutrinos, tectonic plates, or even genes.

---

[16] Of course I would need to say what I mean by 'verification' and 'independent confirmation'.

§1 Introduction: The Realist Turn

The essential provisionality of scientific theories seems to have been forgotten. I recall Fred Hoyle and Hermann Bondi holding centre stage with their Steady State theory of the universe in which there was no Big Bang and matter was continuously created to compensate for the expanding nature of the universe, thus keeping the average density constant. This was the standard accepted story. Is it really appropriate for philosophers to so quickly accept a completely different theory as if it was finally agreed fact[17] – part of that 'completed physics' that realist philosophers are so apt to speak about?[18]

Moreover, if next week the Big Bang is discarded and science adopts a new theory,[19] must philosophers immediately follow them? Here I recall van Fraassen's rejection of the materialist/physicalist stance that looks to science for its *ontology*, in favour of the empirical stance that looks to science for its *method* (2002: §2).

We do not need to embrace an extreme Mach-like instrumentalist anti-realism, nor even van Fraassen's gentler agnosticism, as *tout court* rejections of realism, but rather we

---

[17] One could cite many philosophers who do this: Frank Jackson (2003) adds a gratuitous factual Big Bang to "Russell's hypothesis": 'the world came into existence five minutes ago containing each and every putative trace that might suggest that it has existed since the big bang'. Colin McGinn (1995), discussing the origin of consciousness, speaks as if the Big Bang were an established *fact*, rather than a fallible scientific theory, claiming that: 'it was at the moment of the big bang that space itself came into existence'.

[18] Many works refer to a 'completed physics', including Jackson (1991: p291); Brian Loewer (1996: p103).

[19] A possibility – Andrei Linde's 'fractal inflation' postulates an eternally existing universe, small regions of which undergo cosmological-scale inflation as the result of random quantum fluctuations. Some string theorists claim that their models preclude a Big Bang style singularity. I take no position as to the correctness or otherwise of any scientific theory, mentioning this here solely as a reminder that any theory can be revised or replaced and should not be taken as revealing *established fact*.

should say that the epistemic bar has been set too low; that in the realist turn ushered in by Sellars, Smart, et al, the pendulum has swung too far in the opposite direction. We have moved from an extreme anti-realism to an equally extreme naïve realism in which science is seen as an unfailing (and unchanging) oracle concerning ontology. Neither extreme was correct.

Broadly speaking I favour an empiricist viewpoint such as that which van Fraassen has brilliantly expounded over the last thirty years[20] with an admixture of the pragmatist views of Larry Laudan.[21] This would involve the adoption of a stance or attitude based around something like the following tenets:

i) <u>Knowledge</u>: A preference for knowledge based upon our experience of the world – empirical observation that is verifiable in experience.

ii) <u>Explanation</u>: A rejection of the view that the immediate objects and phenomena of experience stand in need of explanation in terms of further items not directly available to experience. This includes a distrust of explanation by postulation.[22] My own methodological rule would be: Posit nothing as an explanation that would be beyond any possible test of experience.

---

[20] See, for example, his 1980, 1985, 1989, 1994, 2002, 2007.
[21] See, for example, his 1977, 1981, 1983, 1984b, 1990a, 1996. However, as will be seen, there are certainly issues on which I disagree with both Laudan and van Fraassen.
[22] This corresponds to van Fraassen (2002: p37):
(a) a rejection of demands for explanation at certain crucial points.
(b) a strong dissatisfaction with explanations ... that proceed by postulation.

iii) <u>Methodology</u>: A respect for science, not for what it says exists (the position of both realists and physicalists) but for its methodology (discussed in §9.3).

iv) <u>Rejection of substantive metaphysics</u>: Particularly the kind of metaphysics that claims to follow the methods of science while not exposing itself to the rigorous selection process employed by science (see §9.3). I align myself with the critiques of metaphysics of van Fraassen (2002: §1) and Ladyman & Ross (2007: §1). I am attracted to what Ronald Giere calls 'the strategy of replacing metaphysical doctrines by methodological stances' (2006: p39).

v) <u>Useful prediction, not discovery of truth</u>: A conception of science as concerned with prediction and means of intervention rather than the revealing of a hidden reality, and a belief that the engine of scientific progress has mainly been pragmatic usefulness, rather than a concern for truth.[23]

It is not the aim of this essay to defend these views, scientific anti-realism in general, or van Fraassen's philosophy in particular. Rather, the aim of this essay is to show that the arguments offered by realists are unconvincing and offer no reason why the kind of position I have outlined should not rationally be held.

---

[23] One could quote Peter Lipton's saying of A.J. Ayer that 'Science remained for him at base an instrument for the anticipation of experience' (1994: p89). To use the words of Edward Skidelsky, I favour the 'Baconian image of science as power and manipulation,' and not the 'Galilean conception of science as *theoria*, as vision.' (2008: p12).

§2 What is Scientific Realism?

§2.1 Varieties of Scientific Realism

Having briefly discussed the historical background and the overall objectives of this essay, I must now refine what exactly I refer to by the phrase 'scientific realism'. In order to do this I shall examine the ways in which various philosophers of science have stated what they mean by 'realism' and then distil from these the particular claims I shall use to characterise the position.

Before going further on this, a word about 'entity realism' – a variety of scientific realism that I shall not examine in any detail. Entity realists are committed to the existence of the theoretical items (see glossary) referred to by scientific theories, but they are not committed to the truth of scientific theories, nor to any laws they may entail. As such, their realism is not about theories but about the existence of theoretical items. Thus, for example, the entity realist would claim that Thomson, Millikan, Rutherford, Bohr, and others, all referred to the electron despite having very different theories about it.[1] Entity realists raise important and

---

[1] Although we have the generic term *entity realism*, its adherents do not have much in common, other than commitment to the ontological aspects of science. Nancy Cartwright (1983) is non-realist concerning scientific laws, but her realism concerning theoretical entities is typified by this discussion of the phenomenon of a track in a cloud chamber: 'In explaining the track by the particle, I am saying that the particle causes the track, and that explanation ... has no sense unless one is asserting that the particle in motion brings about, causes, makes, produces, that very track ... If there are no electrons in the cloud chamber, I do not know why the tracks are there.' (1983: pp92,99). Her entity realism is based upon causal considerations, but Ian Hacking argues that the ability to manipulate and use electrons, justifies his belief in their existence, hence his pragmatist slogan 'if you can spray them, then they are real'. (1983: p22). Other philosophers whose realism is restricted to entities include Devitt

§2 What is Scientific Realism

interesting issues, but this essay is primarily concerned with realism/anti-realism in terms of scientific theories and for that reason, as well as considerations of essay size, I shall not discuss entity realism further.

A scientific theory comprises a number of parts. It may include some mathematics, and some logic. It will also include reference to observable items such as location in space, velocity, observed temperature, etc. In addition, the theory will contain theoretical terms that are postulated by the theory.[2] Some of these theoretical terms are to be understood as referring to entities, processes, events, or properties. This is not to prejudge the question of whether these latter actually exist, for that question is, at least in part, what divides realist and anti-realist, so let us at this stage allow that these theoretical terms may or may not successfully refer.[3] What is agreed on all sides is that it is certainly necessary for the success of the theory that what it says about the observable items is borne out by experiment – the theory must 'capture the phenomena'.

Stated in its simplest and briefest way, many scientific realists claim that those theoretical terms that putatively refer, do so successfully – i.e. their referents actually exist and largely conform to the description of them that the theory entails. In a trivial sense that claim would automatically entail that such theories are true – if theoretical terms successfully refer then it is true that they successfully refer, and if they all

───────────────────── footnote continuation

(1997), and Ellis, who says 'If the world behaves as if entities of the kinds postulated by science exist, then the best explanation of this fact is that they really do exist' (1990, p53). But again they do not pursue anything like the same project.

[2] The terms 'theoretical', 'observable', 'unobservable' are discussed in detail in §2.2.

[3] Perhaps the real issue is not whether theoretical terms refer, but *what* they refer to – a mathematical object, something whose 'existence' is relative to the theory, or something that has mind-independent existence.

successfully refer then the theory can be called true. However, it is also notable that most realists want to bring truth more explicitly into the picture, by claiming that theories are approximately true and that successive theories in some sense converge upon the truth. I said that the above is true of *many* realists, but there is another group who avoid claiming truth/approximate truth for theories, preferring to state their realism in terms of the *aim* of science being truth. This difference will become important later, but perhaps we should first make precise the claims of scientific realism, and here we encounter a problem since there are almost as many statements of realism as there are realists, and the differences are often quite major. Appendix 1 examines the formulations of scientific realism from various philosophers of science in terms of their adherence to five different kinds of claim – epistemic, metaphysical, semantic, axiological, and whether or not they include correspondence truth as an essential ingredient of their realism. I have chosen those philosophers whose primary concern is to understand, interpret, and explain science, its activities, and its theories, and not those aiming to establish a wider metaphysics around science.

With that in mind I must qualify my use of the word 'metaphysical'. I am concerned to refute the arguments of the various philosophers of science who propose a realist interpretation of the nature of science, its activities and its theories. I use the word 'metaphysical' to refer to some of their claims because those philosophers themselves use it. However, *qua* 'metaphysical', those claims are of a comparatively modest nature and might better be described as 'ontological' or 'existence' claims – that whatever the theoretical terms of scientific theories refer to, they actually do exist. I say 'modest' in comparison to the much broader claims made by those wishing to establish a deeper substantive metaphysics of science. Here I refer to philosophers working on powers, dispositions, natural kinds, essentialism, etc. Bird, Ellis, and Mumford have already been mentioned in connection with essentialism (p188n12), but there are many others, for example, David Armstrong and E.J. Lowe. Such work is not primarily aimed at giving an interpretation of science, its theories and activities, but is concerned to establish a generally applicable metaphysics which, those philosophers believe, derives from the findings of science, and explains science being what it is. I am inclined to describe these philosophers as working primarily within the philosophical discipline of metaphysics, and not the discipline of philosophy of science, but that might be a contentious distinction.[4] Consideration of this substantive metaphysical work is beyond the scope of this essay, though my remarks in §1.3 make it clear that I would be critical of its ambitions. Henceforth, when I use the word 'metaphysical' it should be understood in this more modest sense.

Here are examples of those four kinds of claim – epistemic, metaphysical, semantic, axiological, and I shall refine them later:

Epistemic: Mature and predictively successful scientific theories[5] are well-confirmed and approximately true.

---

[4] Some, such as Bird, inhabit both disciplines.

[5] In line with the first two conditions listed by Laudan (1983: p89), I take a scientific theory being *successful*, to mean that it enables us to make more correct predictions than we would without it. This can be taken as implying that it enables successful interventions in the natural world. Such a theory has *predictive success*.

Metaphysical: The unobservables of well-established current scientific theories exist and have most of the properties attributed to them by science.

Semantic: Theoretical terms in scientific theories should be thought of as putatively referring.

Axiological: A scientific inquiry aims to discover the truth.

Table 1 summarises the findings of Appendix 1 and shows the variety of preferences of the various writers I have considered here. I stress that it is not intended to capture each writer's total view, but rather, specific formulations of realism they have made. I include the formulations of two people who would not normally be thought of as realists – van Fraassen and Ladyman[6] – as well as the proposals of those who would call themselves realists.

---

[6] Ladyman is no anti-realist, but with his commitment to *Ontic Structural Realism* (see his 1998 & 2002), it is misleading to think of him as a realist within the terms of this discussion.

§2 What is Scientific Realism

	Metaphysical	Semantic	Epistemic	Axiological	Correspondence Truth
Boyd	✓	✓	✓		
Chakravartty	✓	✓	✓		
Devitt	✓				(n1)
Sankey	✓	✓	✓	✓	✓
Ladyman		✓	✓		
Musgrave		✓		✓	✓
Niiniluoto	✓	✓	✓	✓	✓
Psillos	✓ (n2)	✓	✓		✓
Sellars	✓		✓		
van Fraassen			✓	✓	(n1)

Table 1: The Varieties of Scientific Realism

Notes:
n1: Although not part of their statement of realism, both advocate correspondence truth.[7]
n2: His metaphysical claim is non-standard, referring to natural kinds.

One thing is immediately noticeable, and that is that realists and non-realists seem to be talking past each other, for the former show a degree of agreement about the need for a metaphysical claim, but both of those I have called non-

---

[7] See van Fraassen (1980: p197), Devitt (1997: pviii). Recently van Fraassen's position has changed as he has realised that *correspondence truth* gives him some metaphysical entanglements he would rather avoid. He now favours a deflationist account (2006b: p153, 2008: pp247-9).

realists – van Fraassen and Ladyman – concentrate on the epistemic claim and ignore the metaphysical one.

However, before proceeding further I need to discuss the terms 'observable', 'unobservable', and 'theoretical', partly because, as Appendix 1 shows, it is impossible to discuss realist claims without immediately encountering these terms. I shall also discuss the observable/unobservable distinction as it affects the realism/anti-realism debate.

§2.2 Observable, Unobservable, Theoretical, and Some Distinctions

§2.2.1 Observable

It is difficult to give a simple and uncontentious definition of 'observable'. Nevertheless, let me begin with van Fraassen's:

> X is observable if there are circumstances which are such that, if X is present to us under those circumstances, then we observe it. (1980: p16)

He also says that 'A flying horse is observable – that is why we are so sure that there aren't any ...' (*ibid:* p15). An eclipse is a phenomenon that is *observable* – how could it not be? Humans located at the appropriate place and time will agree to have observed an eclipse, even if that fact is relative to the word 'eclipse' being theory-laden. van Fraassen himself draws out the distinction between *observing* and *observing that* (*ibid:* p15). Our humans who observe an eclipse may fail to *observe that* it is an eclipse, but even if the theory-laden word 'eclipse' is discarded, those humans will give various descriptions of what they have observed to which we would attach the name 'eclipse'. Similarly, the height reached by a rocket is an observable quantity, as is a pointer reading on a dial, or a thermometer reading. Conversely, no human will ever agree to having observed a quark, unless the word 'observed' changes its meaning, perhaps by becoming equated to the different word 'detected'. As van Fraassen observes, the human organism is 'a certain kind of measuring apparatus' (*ibid:* p17) and in that respect *could* be said to form part of an experiment for the detection of quarks. From there it would be a short step to saying that humans detect quarks.

§2.2.2 Unobservable

Turning to the word 'unobservable', the theoretical terms of scientific theories may refer to putatively existing items that are beyond the reach of human observation, either in principle, for example, quarks or photons, or because of epistemic limitations in our sensory apparatus, for example, microbes.[8] All of these items are beyond direct epistemic access by human perception, hence unobservable. Here the distinction between observe and detect is central since, as van Fraassen remarks of particle trails in a cloud chamber, 'while the particle is detected by means of the cloud chamber, and the detection is based on observation, it is clearly not a case of the particle's being observed' (*ibid:* p17). Detection of a particle in a cloud chamber does not make that particle observable.

§2.2.3 The Observable/Unobservable Distinction

Much ink has been spilled concerning the validity of the distinction between observable and unobservable items (entities, events, or processes), but in this essay I shall largely take it for granted since it would otherwise be impossible to engage with the statements of realists. I will not mount an extended defence of the distinction, but nevertheless, some

---

[8] Assuming that the distinction between 'observe' and 'detect' is retained.

remarks are in order. The distinction is nothing to do with scientific theories, but is an epistemic distinction applying not to the terms of a theory but to entities, events, or processes and their relation to human observers. This latter qualification already suggests that we may anticipate some vagueness.

Churchland (1985) attempts to invalidate the distinction by using a thought experiment concerning the possibility of humans with electron microscopes in place of eyes. However, this seems irrelevant because the distinction is already acknowledged to be relative to the capabilities of us human observers. If human capabilities changed then the point where the distinction is drawn would change, just as the point where we call a man tall has changed as the average population height has increased.

The question of which side of this observable/unobservable divide we should place items perceived through glass microscopes is contentious. van Fraassen thinks such items are not observable but he concedes that some may disagree.[9] However, none of this undermines the validity of the distinction, merely confirming what we expected – that it is a vague distinction, but a vague distinction is still a distinction. van Fraassen again:

> A vague predicate is usable provided it has clear cases and clear counter-cases. ... A look through a telescope at the moons of Jupiter seems to me a clear case of observation, since astronauts will no doubt be able to see them as well from close up. But the purported observation of micro-particles in a cloud chamber seems to me a clearly different case. (1980: pp16-17)

---

[9] For example: 'The main points of our discussion are not much affected by just where precisely the line is drawn' (2008: p110). He goes on to say that he draws the line 'this side of things appearing in optical microscopes', but would not mind if others drew it this side of the electron microscope.

van Fraassen's defence of the distinction is well known, and if this essay was concerned primarily with setting forth an argument for anti-realism then I would need to discuss that defence and the arguments that have been levelled against it.[10] However, my purpose is to undermine the claims of scientific realism, and I will not defend the distinction further as I think there are (at least) two fundamental reasons why I am entitled to simply take it for granted.

Firstly, the absence of such a distinction would render meaningless great swathes of writing by both realists and anti-realists. To give just one example: it would not be possible to formulate one of the most important problems in philosophy of science – underdetermination – if there was no distinction at all between observable and unobservable. For example, is the phrase 'underdetermination of theory by evidence' even meaningful without the distinction? A casual inspection of almost any work on underdetermination will show references to observable and unobservable littered throughout the text. Take, for example, this discussion by Michael Devitt:

> There are two dimensions to the realisms that are challenged by underdetermination. "The existence dimension" of Commonsense Realism is a commitment to the existence of most observables such as stones, trees, and cats ... The existence dimension of Scientific Realism is a similar commitment about such unobservables as atoms, viruses, and photons. (2002: p26)

---

[10] For example, that of Ladyman (2000), replied to by Monton & van Fraassen (2003), and pressed again by Ladyman (2004). Ladyman's objection is not specifically against the observable/unobservable distinction but against van Fraassen's use of it in view of Ladyman's claim that it commits van Fraassen to a metaphysics which he claims to abjure.

Or this by Boyd:

> Call two theories empirically equivalent just in case exactly the same conclusions about observable phenomena can be deduced from each. Let T be any theory which posits unobservable phenomena. (2002: §2)

Both Devitt and Boyd are scientific realists, but these examples are hardly surprising since the very notion of underdetermination involves the idea that evidence underdetermines those parts of a theory that go beyond the evidence. One suspects that it may be hard to uphold the distinction between that which is entailed by the evidence and that which 'goes beyond the evidence' without reference to some kind of observable/unobservable distinction. Regardless of how difficult and contentious the distinction may be, much of the philosophy of science would need rewriting without it.

The second reason why I don't believe it is incumbent upon me to defend the distinction is this: As we shall see in the rest of this chapter, and as Appendix 1 shows, not only do most realists freely use the distinction in discussing underdetermination, they also rely upon it for the formulations of their realist positions. Most of the debate in which I aim to undermine the arguments of realism depends on the distinction. I offer two examples of this dependence, both from the writing of prominent scientific realists. Firstly, this recent definition of scientific realism by Devitt:[11]

> Most of the essential unobservables of well-established current scientific theories exist mind-independently... (2004: p102)

---

[11] We will encounter this again in §2.3.1 (and see also Appendix 1).

My second example is from David Papineau's introduction to his (1996) collection of essays. Almost immediately, in a section titled 'Realism and its antitheses' that aims to characterise scientific realism, he refers to: 'claims about unobservable objects like electrons' (p4). He continues, 'nearly all contemporary philosophers of science accept that science aims literally to describe an unobservable world ...' (*ibid.*). Yet he gives no prior definition of 'unobservable', a word he clearly assumes he is entitled to take for granted.

If one's aim is to undermine a philosophical position, it is surely reasonable to accept the way that position is formulated by those who hold it! As we shall see later, without the observable/unobservable distinction it seems that a particular kind of realism cannot even be stated. It thus seems justifiable to take realist claims at face value, including their assumption of this distinction.

Some protest that it is only the anti-realist who is truly reliant on the distinction as it is essential to his position. This objection goes back to Grover Maxwell (1962), and is also voiced by Alan Musgrave (1985: p204) who says 'Anti-realists need to draw a dichotomy between theory and observation'. More recently, André Kukla has suggested that difficulties surrounding the distinction constitute an argument against anti-realism. Believing the distinction to be of greater importance for the anti-realist than for the realist, he claims that 'the greater dependence of anti-realism on the distinction is intuitively compelling', and that without a theory-observation distinction 'there can *be* no anti-realism'.[12] I do not share his intuition, which seems to be shared only by some realists, which alone gives reason for suspicion.

---

[12] See his 1998: p111. Nevertheless, it is notable that after three chapters of detailed discussion, even Kukla grants a valid way of drawing the distinction (1998: ch.11).

Moreover, if the distinction is not central to scientific realism why do so many formulations of that position rely upon the distinction?[13] Kukla's view almost amounts to the suggestion that scientific realism is an entirely negative thesis that can only be stated as the negation of scientific anti-realism; that without anti-realism, realism could not be formulated as a thesis in its own right. I believe that both scientific realism and anti-realism are independently viable philosophical viewpoints[14] both of which are equally reliant upon the distinction between observable and unobservable.

§2.2.4 Theoretical and Unobservable

The next thing to note is that, as we shall see later, my examination of the claims of actual realists exposes a tendency to run together the words 'unobservable' and 'theoretical' in a manner that makes them seem almost interchangeable. To take just one example – Richard Boyd's formulation of realism:

> The historical progress of mature sciences is largely a matter of successively more accurate approximations to the truth about both observable and unobservable phenomena. Later theories typically build upon the (observational and theoretical) knowledge embodied in previous theories. (see Appendix A1.1)

In the first sentence observable and unobservable are contrasting terms, whereas in the second sentence observational is contrasted with theoretical. There seems, in effect, to be an assumption that all theoretical terms of a scientific theory are necessarily intended to refer to

---

[13] Some, such as Musgrave, proposing an axiological realism, claim not to rely upon the distinction, but I shall later suggest that in fact they do (see §3.7.2).

[14] Or perhaps *stances*, to use van Fraassen's (2002) terminology.

unobservables; that theoretical entails unobservable. This way of understanding the word theoretical seems to assume that it must always refer to that 'hidden reality' the existence of which the anti-realist contests. In his discussion of these issues van Fraassen seems to take the same position, suggesting that he too thinks of the theoretical as *eo ipso* unobservable. He rightly says that two different distinctions are sometimes confused – that between theoretical and non-theoretical, and that between observable and unobservable:

> Terms or concepts are theoretical (introduced or adapted for the purposes of theory construction); entities are observable or unobservable. This ... separates the discussion into two issues. Can we divide our language into a theoretical and non-theoretical part? On the other hand, can we classify objects and events into observable and unobservable ones? (1980: p14)

He goes on to reject the theoretical/non-theoretical distinction on the grounds that 'All our language is thoroughly theory infected' (p14). This does seem to be a widely held understanding of what 'theoretical' means in a scientific context.

As someone joining a pre-existing debate, I will retain the standard meaning of the word, but there is a different way to understand the word 'theoretical' that is not automatically tied to the unobservable, as this example illustrates: in 1843 Adams and Leverrier independently proposed the theory that various discrepancies in the predictions of planetary behaviour could be accounted for by the postulation of a new planet, let us call it p. It could be claimed that p was a theoretical term in this theory, and that it referred to a putative planet. Later 'p' was discovered and named Uranus but even before its discovery it was known to be observable, for while a planet might be as yet unobserved, it could hardly be unobservable.

Construing 'theoretical' in this way, we are not forced to assume that the referent of a theoretical term is necessarily unobservable. Rather, a theoretical term would refer to something the theory postulates, as opposed to other terms that are simply taken for granted, or inherited from other theories.[15] So it would now be possible to concede that all language is theory laden but still draw a useful theoretical/non-theoretical distinction concerning the terms of scientific theories. I will not set out an extended defence of this view here because it is of only peripheral concern to this essay. Nevertheless, the following comments might indicate the way such a defence would go: van Fraassen concedes that the way scientists talk 'is guided by the pictures provided by previously accepted theories.' This seems to leave open the possibility that simply because various terms of a theory are theory-infected in terms of other theories, does not contradict the view that within any particular theory a distinction between theoretical and non-theoretical can be drawn. Indeed, it seems to me that it is not possible to make sense of scientific theories unless this is conceded. Suppose some new theory is devised to account for what Quantum Mechanics currently accounts for. Such a theory involves the observable phenomena, and let us concede that references to this are all 'theory infected' by the language of other theories. The new theory would postulate the existence of new entities via the use of new theoretical terms. There would clearly then be a distinction between these new 'theoretical entities' and the other theory-laden entities of the theory, the distinction being that the theory postulates the former but takes for granted the

---

[15] Some theoretical terms are not intended to refer, for example the term 'inertial frame'.

§2 What is Scientific Realism

latter. This is just what is found in the case of the discovery of Uranus. On the one hand we have the set of facts that the new theory confronts – the theory infected phenomena – and on the other hand we have the theoretical item(s) postulated by the theory. The fact that all observation is theory-laden need not prevent the drawing of a distinction between posited theoretical and pre-defined non-theoretical terms within any given theory.

Whether or not an entity, event, or process is observable is an epistemic matter concerning the relation between it and us human observers. However, whether or not a term in a scientific theory is 'theoretical' is decided in a completely different manner bearing no relation to observability, but being more connected with the role played by that term in the theory. A better distinction to draw might be between theoretical as posited and non-theoretical as predefined. Moreover, this would have the advantage of not having to concentrate on the observable/unobservable distinction. After all, should it be necessary to get caught up in the arguments over the observable/unobservable *epistemic* distinction in order to debate scientific theories and whether the entities they postulate do actually exist? Perhaps an epistemic question belonging to the philosophy of perception should not take such a central place within the realism/anti-realism debate.

Nevertheless, since I am concerned to undermine realist claims, I am obliged to stick with the more general usage of the word 'theoretical' which, as I have argued, seems to equate to unobservable. This fact further reveals the realist's dependence upon the observable/unobservable distinction because she seems unable to define 'theoretical' other than in terms of 'unobservable'. For if there is no observable/unobservable distinction, and if the theoretical just *is* the unobservable, then what can the realist mean by the word 'theoretical'?

§2.3 Refining the Realist Claims

I now return to the task of establishing precisely what are the claims of scientific realism. However, given the plethora of formulations it is not easy to decide upon the formulation that fairly characterises both what I oppose and what actual realists support – I don't wish to attack a straw man. How does one compare a statement couched purely in metaphysical terms with another couched only in epistemic terms, or yet another in axiological terms? Some of this diversity of expression arises from the fact that there is much genuine variation among the different positions being advocated, variation stemming from the very different degrees of emphasis placed on the four realist themes – metaphysics, epistemology, semantics, and axiology, or five if one includes the correspondence truth claim. I will eventually distil these various claims down to two main positions that I aim to critique.

I will now give a short list of proposed formulations for each of those four claims – semantic, epistemic, metaphysical, and axiological. These are all extracted from the actual claims discussed in Appendix 1. In each case I then select the formulation I intend to take as most representative.

§2.3.1 Metaphysical

a) The reality which scientific theories describe is largely independent of our thoughts or theoretical commitments.

b) Most of the essential unobservables of well-established current scientific theories exist mind-independently and have most of the properties attributed to them by science.[16]

Here (a) seems a watered down version of (b) and the latter thus seems the better formulation. However, in this form it refers to unobservables, but this can be avoided thus:

c) Most of the entities referred to by the theoretical terms of well-established current scientific theories exist mind-independently and have most of the properties attributed to them by science.

§2.3.2 Semantic

a) Theoretical terms in scientific theories should be thought of as putatively referring expressions.

b) The theoretical terms featuring in theories have putative factual reference.

c) A realist semantics implies that a theory's theoretical claims have truth values, and should be construed literally, whether true or false.

d) The observational and theoretical terms within the theories of a mature science genuinely refer.

---

[16] As discussed in §2.2.4, note that these two claims display an apparent assumption of equivalence between *unobservables* and *theoretical commitments*.

e) Scientific discourse about theoretical entities is to be interpreted in literal fashion as discourse that is genuinely committed to the existence of real unobservable entities.

(a), (b), (c) and (e) say essentially the same thing – that scientific theories should be understood such that their theoretical terms are not simply instrumentally useful fictions, but putatively referring. This is not the same as saying that they do successfully refer, but that theories should be read as if these terms are capable of referring. A theory as a whole is therefore capable of being true. (d) seems to go too far in insisting that they actually *do* refer. I propose to adopt (a) as representative of these claims.

§2.3.3 Epistemic

a) Mature and predictively successful scientific theories are well-confirmed and approximately true of the world.
b) The claims of scientific theories give us knowledge of the world.
c) Scientific theories are typically approximately true and more recent theories are closer to the truth than older theories in the same domain.
d) The historical progress of mature sciences is largely a matter of successively more accurate approximations to the truth about both observable and unobservable phenomena.

I think that (c) seems the best formulation here since it includes the convergence claim and the approximate truth claim, both of which I see as central to scientific realist thought.

§2.3.4 Axiological

Here are the formulations of van Fraassen, Musgrave, Niiniluoto, and Sankey:

a) Science aims to give us, in its theories, a literally true story of what the world is like
b) The aim of a scientific inquiry is to discover the truth about the matter inquired into.
c) Truth is an essential aim of science
d) The aim of science is to discover the truth about the world

The axiological claim requires considerable analysis and discussion and is the subject of an entire chapter (§3) so there is little point in attempting to simplify these claims at this point, and I will not do so, though I shall treat Musgrave's (b) as provisionally representative for now.

§2.3.5 Summary of Realist Claims

So we end up with these formulations of the five realist claims:

RC1 Most of the entities referred to by the theoretical terms of well-established current scientific theories exist mind-independently and have most of the properties attributed to them by science.

RC2 Scientific theories are typically approximately true and more recent theories are closer to the truth than older theories in the same domain.

RC3 Theoretical terms in scientific theories should be thought of as putatively referring expressions.

RC4 The aim of a scientific inquiry is to discover the truth about the matter inquired into.

RC5 The correspondence truth claim – that the correct way of understanding truth is via the correspondence theory.

RC1 is the metaphysical claim – that the referents of those theoretical terms do, in general, exist, or at least, acceptance of scientific theories entails acceptance of the existence of the theoretical items referred to by those theories.

RC2 is the epistemic claim – successive theories are converging upon 'the truth'. That convergence upon truth also captures the realist conception of scientific progress.

RC3 is the reference claim – the theoretical terms of scientific theories are to be understood as putatively referring.

RC4 is the (provisional) axiological claim, a statement of what scientific theories aim to achieve. I postpone further discussion of this until §3.

§2.4 Discussion of Realist Claims

Before considering the specific claims let me discuss the centrality to scientific realism of claims I have classified as 'metaphysical'. As we have seen, entity realists such as Devitt and Ellis lay stress on realism being seen as a purely metaphysical thesis. Devitt claims that the epistemic and semantic claims of some formulations of scientific realism are parasitic upon the metaphysical one.[17] This does seem at least plausible – it seems hard to imagine any realist not being concerned with the question of what actually exists.

---

[17] See Appendix 1.

§2 What is Scientific Realism

However, unlike Devitt, I do not think this means that the epistemic and semantic claims can be set aside as unimportant. Notwithstanding the claims of some entity realists, I think it is impossible to claim that the theoretical items of a scientific theory exist without at the same time discussing what that theory says about them, how it refers to them. For example, if one claims 'electrons exist' it is hard to see how this could even have any determinate meaning absent some reference to what the theory *says* about electrons. For without those statements made by the theory it is unclear what the term 'electron' means. It seems to be just an empty gesturing toward a *something* that has some causal impact, but as soon as anything is stated *about* the electron, semantics becomes involved.[18] Similarly, how can we claim the existence of some theoretical item without any discussion concerning how we come to know of its existence – what is the justification for the existence claim? Here my latent verificationist thinking may be showing, but I see little sense to be made in claims for the existence of something absent at least some indication of how we come to know about it. Conversely, what sense can be made of how we come to know about something without first assuming that it exists? I do not need to lose completely the gap between existence and justification in order to suggest that claims concerning both are linked.

---

[18] Let me recall Musgrave's comment, which illustrates the tendency to vacuousness of the ontological claim unless accompanied by some kind of descriptive referential claims: 'We are to believe in scientific entities ... without thinking true any theory about those entities ... This is incoherent. To believe in an entity, while believing nothing further about that entity, is to believe nothing. I tell you that I believe in hobgoblins (believe that the term 'hobgoblin' is a referring term). So, you reply, you think there are little people who creep into houses at night and do the housework. Oh no, say I, I do not believe that hobgoblins do that. Actually, I have no beliefs at all about what hobgoblins do or what they are like. I just believe *in* them.' (1996: p20)

Moreover, just as Devitt argues that the metaphysical claim of realism entails some kind of corresponding semantic and epistemic claims and that these can therefore be dispensed with, one could equally argue that the semantic and epistemic claims entail a corresponding metaphysical claim that could similarly be dispensed with. Metaphysics, semantics, and epistemology seem to jostle for superiority here. Consider these examples of putative realist semantic and epistemic claims:

R1 Scientific theories ... are typically approximately true and more recent theories are closer to the truth than older theories in the same domain

R2 The observational and theoretical terms within the theories of a mature science genuinely refer. (Laudan, 1981: p20)

Clearly no realist could believe these claims without also believing a corresponding metaphysical claim.[19] Even van Fraassen's explicitly axiological/epistemic formulation of realism can be argued to have a metaphysical implication:

> Science aims to give us, in its theories, a literally true story of what the world is like: and acceptance of a scientific theory involves the belief that it is true. (1980: p8)

A realist who accepts a theory and believes it true, must surely believe that its theoretical terms refer to actually

---

[19] Laudan goes on to say that R2 roughly corresponds to 'there are substances in the world that correspond to the ontologies presumed by our best theories' (p20), thus converting what ought to be a purely semantic claim concerning how theories should be interpreted, into what is, in effect, a metaphysical claim masquerading as a semantic one.

existent items – for that is what the realist means by claiming that the theory is true. Thus such a realist implicitly makes a metaphysical claim.[20] I conclude that, to a considerable extent, realist claims of a metaphysical, epistemic, and semantic form cannot be kept apart since each kind of claim entails claims of the other kinds. Notwithstanding that, I now give reasons for largely excluding from further consideration claims RC3 and RC5.

First, RC3: since RC1 claims that the theoretical items that a theory names do actually exist, that would certainly ensure that RC3 must be true. So RC1 is sufficient for RC3, but it is not necessary since the intention to refer expressed by RC3 might be unsuccessful.

I do not intend to argue against RC3, and I accept it within the scope of this essay. I do not take the 'should' in its formulation as expressing a view as to what various scientists think or intend, but rather that the correct way to read and understand scientific theories is that their theoretical terms are capable of referring, regardless of whether they in fact succeed in doing so. It follows from this that a theory as a whole is capable of being true. Note that in accepting RC3 I do not thereby accept any variety of realism. A realist claim consisting only of RC3 would be far too weak, and all formulations of realism that do use RC3 conjoin it to one of RC1, RC2, RC4. I think that RC3 does rule out some instrumentalist versions of anti-realism, but in other respects it seems that RC3 is actually neutral regarding realism or anti-

---

[20] van Fraassen himself might disagree since he has tried to explicate what it is to *accept* a theory in terms which do not entail belief. For example, he says that 'acceptance of a theory may properly involve something less (or other) than belief that it is true' (1980: p9). However, what I am discussing here are the claims of realists, and no realist could agree with van Fraassen's understanding of *theory acceptance* on pain of, in effect, accepting *constructive empiricism* itself.

realism since many other anti-realists could agree with it, as van Fraassen demonstrates:

> After deciding that the language of science must be literally understood, we can still say that there is no need to believe good theories to be true, nor to believe *ipso facto* that the entities they postulate are real. (1980: p11)

Second, RC5, the correspondence truth claim: even realists are divided about what they mean by the word 'truth'. One might at least expect to find a common allegiance to a non-epistemic account of truth, yet even that is not the case as Ellis explicitly favours an epistemic account.[21] I do believe that a more general concept of theory-world correspondence is of central importance for scientific realism and I discuss this in §8.5. Nevertheless, whilst it does occur in the views of Musgrave, Niiniluoto, and Psillos, I shall not take the correspondence account of truth to be, of itself, a central feature of the formulation of scientific realism, though, as we shall see, the majority of prominent realists are explicit in their allegiance to it. I take this step in line with my wish to reduce to a minimum the claims of realism, and in order to avoid this essay being taken over by the considerable discussion that could be had concerning theories of truth. It seems obvious to me that if I succeeded in providing a definitive refutation of the correspondence theory of truth, I would not thereby have refuted scientific realism. Clearly Devitt (1997: pp43) agrees when he says 'no doctrine of truth is in any way constitutive of realism.'

---

[21] See his 1990, and also his 1985: §V which includes: 'my concept of truth is similar in some ways to the coherence concept defended by some idealists' (p70).

I propose to take RC1 and RC2 as the core claims of a variety of realism that is often referred to in the literature as 'Epistemic Scientific Realism', though I think this is a misleading term. I believe that the term was coined by Laudan (1981) and it was, I suggest, an error even then, for examination of his claims R1-R5 (see this document, p124) show that the word 'genuine' in his R2 clearly makes it an ontological claim. Kukla also uses the term but with a little more justification since he describes 'epistemic realism' as the thesis 'that we can come to know that theoretical entities exist' (1998: p9). However, here it is again hard to see how this claim could be made without there being a corresponding ontological claim, indeed Kukla's claim seems to involve the ontological claim 'that theoretical entities exist'.[22] I will not use the term Epistemic Scientific Realism to describe the position I criticise; instead I will use the admittedly ungainly term Convergent Ontological Scientific Realism (COSR) as I think it accurately reflects the conjunction of the two claims, RC1 and RC2, which I repeat here:

COSR:

RC1 Most of the entities referred to by the theoretical terms of well-established current scientific theories exist mind-independently and have most of the properties attributed to them by science.

RC2 Scientific theories are typically approximately true and more recent theories are closer to the truth than older theories in the same domain.

Having rejected the use of the term Epistemic Scientific Realism, the question nevertheless arises – are some realists more aptly described by that name? We have seen that the

---

[22] Psillos also misuses the term, as I note on p332n2.

great majority take a position that is distinctly ontological, but Clark & Lyons (2002: px) define *epistemic realism* as the claim 'we can be justified in believing that successful scientific theories are (approximately) true'. However, this claim again has an ontological entailment and would stand or fall with claim RC1. Jarrett Leplin makes a very weak claim, which he aptly calls Minimal Epistemic Realism: 'there are possible empirical conditions that would warrant attributing some measure of truth to theories – not merely to their observable consequences, but to theories themselves.' (1997: p103) This claim is genuinely independent of COSR. Space does not permit a detailed analysis of it, but I note that Leplin's book has three out of its seven chapters (2, 3, 5) devoted to the novel predictions argument that I criticise in §6.5 and my arguments there would tell equally against Leplin's position. However, a more fundamental point is that Leplin's claim is so weak that I am not sure I wish to oppose it anyway, for all it amounts to is the claim that there may be at least one instance where we have reason to believe that the theoretical part of some theory has 'some measure of truth'. Some anti-realists might want to oppose even that. However, I make no anti-realist argument in this essay, but merely critique scientific realist arguments of a more general nature than Leplin's. I conclude that my arguments against COSR can be taken as explicitly applying to reasonably strong epistemic formulations of scientific realism, leaving only this extremely weak epistemic claim not specifically discussed.

At this point I must recognise a second quite different formulation of scientific realism, for the axiological claim, RC4, is of a very different kind from the claims of COSR. The latter concern scientific theories and their relation to the world, but RC4 is a very different claim about the intentional attitudes of scientists or science as a whole. I shall refer to the separate claim RC4 as Axiological Scientific Realism (sometimes referred to as ASR). Note that COSR and ASR are not mutually exclusive positions, it being perfectly possible to believe both. COSR is the more important of the two positions as the majority of scientific realists take this position rather than ASR.

I repeat that I do not intend to argue for an anti-realist position, but will rather express scepticism concerning the claims of these two main varieties of scientific realism – COSR and ASR – and I shall argue that there is no rational compulsion to accept them.

§2.5 Plan of Action

In the next chapter I will discuss ASR in detail, and the various kinds of scientific axiological claim. The argument against ASR is quite complex, firstly suggesting that scientific axiological claims in general are of doubtful value and are not what they appear to be. I then offer several arguments against the specifically realist axiological claims, and conclude that they can only survive at all if they are allowed to reduce to claims analogous to those of COSR itself.

The rest of the essay is entirely devoted to COSR and from §4 onwards, whenever I refer to 'realism' or 'the realist' I can generally be taken to be referring to the COSR variety.

§4 will argue that the term 'approximate truth' lacks sufficient definition to enable it to be used as centrally as the realist wishes. I will argue that no definition of approximate truth exists that enables one to decide if RC2 is correct, likely, or even meaningful. I will thus reject RC2 altogether on the grounds that it lacks determinate meaning and thus cannot even get as far as being false.

§5 will present the Pessimistic Induction (PI) argument against realism. I will suggest that, despite its title, it is not best seen as an inductive argument, but as a deductive refutation of the presumption of a link from theory success to theory truth or approximate truth.

§6 will attempt to rebut the large number of realist arguments that have been levelled against the PI (it is long because there are many!) By the end of chapter 6 I believe that the PI will stand as a conclusive refutation of the NMA.

§7 will present an alternative to the usual underdetermination argument against realism which will increase the motivation for scepticism. The argument will be that the historical record suggests that for each of our current successful theories, the existence of unconceived alternatives with different theoretical proposals is highly likely. Consequently, we should be sceptical as to the truth, or even approximate truth, of the theoretical proposals of our current theories. This specifically undermines RC1, and RC2 indirectly so.

§8 will present the No Miracles Argument and lay out the arguments that refute it and show it to be unconvincing. These include a discussion of IBE and the realist's mistaken understanding of its use in science, and also a discussion of the realist's notion of theory-world correspondence and its implicit reliance on IBE.

Finally, assuming that the NMA is refuted and truth is not the best explanation of the success of science, §9 will present an alternative selectionist explanation of that success.

§3 Axiological Scientific Realism (ASR)

§3.1 Introduction

In this chapter I will first examine axiological claims in general, both realist and anti-realist, as they are found within the philosophy of science, and I will argue that they are of dubious value, saying more about the beliefs of the philosophers who make such claims than they do about science. I then criticise the axiological claims of actual realists on several grounds, concluding that philosophers who make such claims are forced to adopt the COSR position, that most current scientific theories are true. I take Musgrave's realist axiological claim as being the paradigm example, though I also refer to the claims of Timothy Lyons (2005) and Howard Sankey (2008). My aim is to show that in addition to the more general objections to axiological claims, in the specifically realist context the problems become insurmountable.

We have already encountered the views of Musgrave and Sankey, discussed in A1.5 and 0. Briefly stated, Musgrave (1996: p19) claims that:

> the aim of a scientific inquiry is to discover the truth about the matter inquired into.

Sankey (2008: p13) says:

> the aim of science is to discover the truth about the world, and scientific progress consists in advance toward that aim.[1]

---

[1] For Sankey this is just the first of six 'core doctrines' elaborated in his (2008: pp13-18). It is clear that his position is close to COSR as well as ASR: 'a great deal of truth has already been discovered in at least some areas of science.'

§3 Axiological Scientific Realism (ASR)

According to Lyons (2005: p167):

science seeks true theories and is justified in doing so [2]

Popper must also be mentioned. I will return to this later (p68), but he claimed 'the aim of science is to explain ...' (1983: p145).

Additionally, in a very recent paper, Bird says 'The aim of science is the generation of scientific knowledge'.[3] Both Musgrave and Lyons claim to be critical of the kind of COSR expressed by RC1 and RC2 and which will be discussed in the rest of this essay. For example, Lyons is critical of what he calls the 'epistemological tenet', which he characterises as 'we can justifiably believe that our successful scientific theories have achieved (or approximated) this goal [of being true]' (2005: p167). And Musgrave (1996: p20) describes as 'mad-dog realism' the view that 'all or most current theoretical entities really exist, that all or most current theoretical terms really do refer', claiming that such a view 'erects current science into a metaphysic and ties scientific realism too closely to that metaphysic' (*ibid:* p21).[4] Thus both writers

---

———————————————— footnote continuation
(p13) and his 'scientific progress consists in advance on truth' (*ibid.*) is reminiscent of my RC2.

[2] As well as 'science seeks', he also says it 'aims' and 'endeavours'. He doesn't say how this should be understood, but there is evidence that his axiological claim refers to the intentional attitudes of scientists. For example, having mentioned Einstein, Newton, and Dalton, he says 'they endeavoured to make that thesis manifest in predictions about the world. Such endeavours ...' (2005: p185). While remaining faithful to an axiological realism, he veers away from truth as the aim of science in favour of a more complex notion – an 'increase in manifest truth' (*ibid:* p177) but I will not pursue that here.

[3] See his (2010: p1), though I will not examine this further except to note his preference for knowledge rather than truth as the correct focus for realists.

[4] Though his discussion of the NMA in his (1985: §III) and (1988) suggests COSR. Indeed, despite his protestations, it would be reasonable to say that he is a COSR theorist concerning those theories that have made novel predictions. I argue against the significance for COSR of novel predictions in §6.5.

claim that it is possible to be a scientific realist without necessarily believing that most current scientific theories are true or approximately true. However, just as some axiological realists are critical of COSR, Psillos, a prime defender of that view, is critical of axiological realism, claiming that 'it seems rather vacuous' (2000: §2.2).

The paradigm example of an anti-realist axiological claim is of course van Fraassen's: 'Science aims to give us theories which are empirically adequate' (1980: p12).[5] There is also Laudan's anti-realist axiological claim that the aim of science is problem solving (1977, 1984b, 1990a) though I will not discuss that further.

I will now examine in some detail the nature of axiological scientific claims in general, as I believe that it is not clear what they mean, and the deeper one searches for clarity, the less useful they become. Such claims typically take the form:

AC1   Science aims to Φ, where Φ is, for example, one of: discover the truth (Musgrave and Sankey) solve problems (e.g. Laudan) develop empirically adequate theories (van Fraassen)

It is true that Musgrave's formulation attributes the aim not to 'science' but to 'a scientific enquiry', but since he clearly believes that all scientific enquiries share the one unitary aim, one is entitled to convert his claim to the AC1 form. These all appear to be claims as to the intentional features, not of scientists, but of science itself, and that in itself raises many questions – as Gideon Rosen says in his discussion of van Fraassen's axiological claims, 'Forces of considerable

---

[5] He also proposes an axiological understanding of scientific realism: 'Science aims to give us, in its theories, a literally true story of what the world is like' (1980: p8).

rhetorical complexity are at work.' (1994: p144) There are different ways to construe the word 'aim' in AC1 and I will first argue that the teleological construal is not viable.

§3.2 Aims Construed as Teleological

Presumably the word 'science' here refers to some abstract or social entity and it is not obvious that such an entity can have *aims* in this teleological sense of the word, these generally being attributable to agents. Doubtless, individual scientists have aims, but that doesn't entail that *science* has aims. I concede that it is reasonable to talk in terms of *some* social entities having aims – for example, the Labour Party, or the Society for the Protection of Birds. However, there are restrictions. To begin with, the social entity needs to possess a reasonable degree of unity. In addition, unless there is some kind of central governing body with an agreed set of rules and objectives, it may be unclear what these 'aims' are, or even if they exist at all in any unified sense. A good example here might be 'medicine' – it might be suggested that this has aims, but does it? If 'medicine' is not qualified it might be taken to refer on a world wide basis, and then it seems unlikely that we shall find the kind of unity required. On the other hand, if we refer only to 'British medicine' we may be inclined to agree on the grounds that British medicine does have such a central body – the BMA – and its aims will be formally written down. However, this can be done successfully only insofar as British medicine can be seen as a unity. As soon as we allow for alternative medicines the claim that British medicine has unitary aims begins to appear problematic.

If we transfer these considerations to what is referred to by the word 'science', two things are pertinent: firstly, there is no single social entity known as 'science'. Secondly, there is

§3 Axiological Scientific Realism (ASR)

certainly no written statement of what the aims of such an entity might be. There are many different sciences and it is doubtful that the word science can act as anything other than a family resemblance term, lacking any implication of shared unitary aims.[6] In contrast, whether realist or anti-realist, the axiological claim AC1 suggests an essentialist conception of science – that there is some *one* feature common to all activities going under that name, whether it be discovery of the truth, or the development of empirically adequate theories.[7] The vision of methodological, reductive, or linguistic unity embodied in the *Encyclopaedia of Unified Sciences* of Carnap and Neurath is surely now a discarded dream and we hardly need Feyerabend to tell us of the radical diversity of the different sciences.[8] The sciences of psychology and geology may well display *some* commonality of method, but it is questionable whether projects within these two sciences share common *aims*.[9] Even if we consider individual sciences as being a possible locus for the word 'aims', the claim still seems doubtful. Consider physics, does it have unitary aims? There is certainly no governing body with a written statement in which we might find such aims. It is also not clear how the word 'physics' is defined. What are

---

[6] As Peter Hacker (1987: p195) says, 'there is no more *one* science than there is *one* language.', a view endorsed by John Dupré who proposes that science is 'a family resemblance concept' (1993: p10), later expanded upon when he says there are 'perhaps an indefinite number of features characteristic of parts of science, and every part of science will have some ... probably none will have all.' (*ibid:* p242).

[7] Recall Niiniluoto's axiological claim: 'Truth is an essential aim of science' (see A1.6).

[8] I already mentioned Dupré; Ian Hacking emphasises a pluralism of 'styles of reasoning' and urges reference to 'the sciences' rather than 'science' as a unity (1996: pp71-74); Cartwright (1999) advocates a pluralist view; the collection edited by Kellert, Longino & Waters (2006) shows much support for such a view; see also Chang (forthcoming).

[9] The possible objection that common method entails common aims is discussed in §3.5.

§3 Axiological Scientific Realism (ASR)

its boundaries? We'd better not say that it is the study of 'the physical' as that will lead us into even deeper problems. Presumably physics is practiced by 'physicists', but who are *they*? There are many physicists and doubtless they disagree amongst each other. Which country's physicists? Academic physicists only, or are their industrial colleagues included? Perhaps people within a university's chemistry or mathematics departments are actually doing physics. Could the physiologist study physics, or the historian of science? The term 'physicist' seems to be inherently vague and therefore, defining 'physics' as that which is done by physicists leaves that word equally vague. This need not mean we must stop talking about anything called 'physics' – a vague boundary is still a boundary. However, the vagueness of the boundary surely makes it impossible for physics (or any of the sciences) to have the unity that is required in order to speak of it having 'aims'.

In the foregoing argument I may be accused of taking for granted one particular position within a major debate within the philosophy of social science – between holism and methodological individualism. However, I would say that I merely abjure the use of teleological claims for science and that those who *do* make such claims need to pay regard to the legitimate arguments of methodological individualism and justify their being set aside.[10] They do not do so.

Of course, individual scientists have teleological aims, and as Rosen says, 'the simplest way to take the claim that science aims ... is as the claim that this is what most scientists aim

---

[10] As presented in Agassi, 1960. More recently, Jon Elster criticises what he calls objective teleology: 'processes guided by a purpose without an intentional subject', a characteristic feature of which is 'to postulate a purpose without a purposive actor' (1982: p454-5). Elster's discussion is in a political context, but is also applicable here.

for.' (1994: p146) However, if the axiological claim is re-interpreted as

AC2   Scientists aim to Φ, where Φ is one of...

then we are no longer in the realm of philosophy but sociology, where some kind of research might be conducted to see if the claim is true.[11] If an opinion survey demonstrates that 99% of all scientists profess to have some specific aim, for example, that of 'uncovering the truth', that could have no impact on a philosophical discussion of science, what it actually achieves, and how it should be understood. If this were not the case then the scientific realism/anti-realism debate could be settled by taking a majority vote among scientists. We can charitably assume this is not what is involved here, this being confirmed by van Fraassen's remark that

> The aim of science is of course not to be identified with the individual scientist's motives. The aim of the game of chess is to checkmate your opponent; but the motive for playing may be fame, gold or glory. (1994: p180) [12]

Perhaps we could take axiological claims as being normative in character, AC1 being then construed as: 'Science (or scientists) *should* aim to Φ, where Φ is ...'. In the completely general case this is a valid possibility, but we will later see

---

[11] As I suggested in p55n2, this may be what Lyons has in mind.
[12] The realist philosopher John O'Leary-Hawthorne (1994: pp128-9) is not so charitable as he construes van Fraassen as making a sociological claim, though he also assumes that scientific realism is partly expressed as a claim concerning the beliefs of scientists.

§3 Axiological Scientific Realism (ASR)

that in the specifically realist case such a construal would be incorrect.

Setting aside that normative construal, the conclusion is that a claim such as AC1, interpreted teleologically, is of doubtful philosophical merit. Interpreted literally it makes unwarranted assumptions as to the unity of science, as well as essentialist assumptions concerning the nature of scientific enquiries. On the other hand, if AC1 is understood as something like AC2 then it is no longer a philosophical claim at all.

In order to avoid these problems, AC1 could be re-written thus:

AC3   Science as an enterprise results in the achieving of Φ, where Φ is one of…

Now this is an important kind of claim since it addresses the question of what the enterprise of science achieves, regardless of what its practitioners or its critics believe, either about that enterprise, or about their own aims. However, AC3 has nothing to do with 'aims' because it is not an axiological claim at all, but a thesis that is entirely expressible as a set of claims concerning what science results in, which could plausibly be taken to be claims about scientific theories and how they should be understood. Such claims would amount to claims similar to RC1, RC2, RC3. Thus any teleological thesis about science seems to reduce to either an empirically testable claim about the intentional attitudes of individual scientists, or to a claim about the theories of science that is itself further reducible to some combination of metaphysical, semantic, or epistemic claims.

Is it possible that AC1 could be reformulated so as to overcome these objections? I have already considered Musgrave's alternative: 'a scientific enquiry aims to Φ', and similar to this would be a claim such as 'a scientific theory aims to Φ'. Although formulations such as these appear to avoid the very wide generality of the AC1 claim, in fact they do not because these apparently reduced claims are intended to apply *tout court* across the whole of science. The most such formulations achieve is to reduce 'Science aims to Φ' to 'All scientific enquiries aim to Φ' or 'All scientific theories aim to Φ', and these formulations are liable to all the objections I have raised.

I therefore believe that formulations of realism in teleological terms are a mistake and I agree with Arthur Fine (1986a: p172) when he urges us 'not to undertake the construction of teleological frameworks in which to set science', and when he further says that 'nothing seems to accrue to our understanding of science if we go looking for general aims or goals' (*ibid:* p174).

§3.3 Aims Construed as Criteria of Success

There is another way of construing the word 'aims' in the axiological claim, and it is the one suggested by van Fraassen when he gives his understanding of both realism and anti-realism in terms of aims. He firstly defines realism thus: 'Science aims to give us, in its theories, a literally true story of what the world is like...' (1980: p8). He then goes on to distinguish the aims of scientists from the aim of science in much the same way as I have done, and finally says 'What the aim is determines what counts as success in the enterprise', giving the example 'The aim of the game of chess is to checkmate your opponent' (*ibid.*). Elsewhere he says:

> if you accept a theory, you must at least be saying that it reaches its aim, i.e. meets the criterion of success in science (whatever that is) (1983: p327)

§3 Axiological Scientific Realism (ASR)

So for him, if science *aims* to achieve true theories, then the criterion of success for science is the construction of scientific theories that are true. Thus, according to van Fraassen, AC1 is equivalent to:

AC4   The criterion of success for science is that it Φ, where Φ is one of: discover the truth (Musgrave and Sankey) solves problems (e.g. Laudan) develops empirically adequate theories (van Fraassen)

Musgrave's claim that 'the aim of a scientific inquiry is to discover the truth about the matter inquired into' (1996: p19) can also be construed as a criterion of success claim.

Note that AC4 refers to '*the* criterion', and not the less contentious '*a* criterion'. van Fraassen concedes that

> in calling something *the* aim, I do not deny that there are other subsidiary aims which may or may not be means to that end ... simplicity, informativeness, predictive power, explanation are (also) virtues (1980: p8)

These other aims are declared by him to be 'subsidiary' though it is not clear what he means by that given that they 'may or may not' subserve what he sees as the primary aim. We shall see later (§3.7.3) that Sankey also concedes the possibility of other aims, but for him they explicitly subserve his claimed primary aim (truth). I shall take AC4 as claiming that although there may be other subsidiary criteria of success, there is just *one* criterion of success common to all the sciences and all of their projects.

AC4 is still an axiological claim of a general kind, and the pluralist objections already raised against the notion of a single unitary science apply equally against this criterion of

§3 Axiological Scientific Realism (ASR)

success conception. Just as there is no common aim, it is equally questionable that there is any criterion of success common to all scientific domains, nor even to all projects within one scientific domain. I concede that any given scientific project will have at least one criterion of success. However, there still seems little to suggest that such a criterion is common to all scientific projects. While it may be at least plausible to claim truth as the criterion of success of the Large Hadron Collider project, it might be more helpful to view the criterion of success of the cancer research project as a successful cure – a successful intervention in the natural order.[13] This example points up the diversity of projects going by the name 'science'. There are indeed theoretical science projects for which truth might at least *seem* like a plausible criterion of success, but much *applied* science blurs into engineering, where Laudan's problem-solving seems a more plausible aim. It would not be helpful to describe Edison's research team searching for the ideal tungsten filament as having truth as a criterion of success, so would that mean that the axiological realist would have to say that it was not a scientific project?

A comment on the pure/applied science distinction: the axiological realist, who holds truth to be the criterion of success throughout science, could try to exclude Edison's work on the grounds that it was not 'pure science', but the topic under discussion is *science*, not *pure science*. Ever since Popper, some philosophers are apt to idealise science in ways that suit their agendas. If it is suggested that Edison's work should be excluded from the domain of science then it could only be on the question-begging ground of assuming that a

---

[13] One can cite an example belonging to the same scientific domain as the LHC project – physics – namely nuclear fusion research, the aim of which is surely cheap and plentiful energy, again via a successful intervention in the natural order.

§3 Axiological Scientific Realism (ASR)

quest for truth is the defining characteristic of science, rather than anything to do with prediction and pragmatic usefulness. That would amount to claiming that Edison's work was not an exception to the rule that science aims for the truth because only pure science is being considered; but the only means offered of demarcating between pure and applied science would be that pure science aims for the truth. This is question-begging. It is similarly hard to see how someone could claim a deep philosophical distinction exists between the LHC project and nuclear fusion research unless they are again relying upon 'science aims for the truth' as that distinction.

Even if one does think of scientific projects as having aims *qua* criteria of success, it seems unlikely that one such criterion would apply to all such projects. As Arthur Fine (1986a: p173) notes, it would be a mistake of logic to go from 'they all have aims' to 'there is an aim they all have', and it would be equally mistaken to go from 'they all have a criterion of success' to 'there is a criterion of success which they all have'. It is ironic that van Fraassen uses the game of chess to exemplify his notion of aim, for games are of course the paradigm example of a group of activities sharing many characteristics but having no common aim, and I have already suggested that 'science' is best thought of as a family resemblance term in the same way as 'game'.

This raises the question of whether there is one single way of describing the aim (*qua* criterion of success) of even a single scientific project. If we consider, for example, the 'Manhattan project', the criterion of success for that project was surely the successful production of a usable atomic bomb.[14]

---

[14] Again, this can't be excluded on the grounds of not being 'pure science'. The issue of pure versus applied science is too large to go into here, but regardless of the pros and cons of that debate, no demarcation can be used that relies on the notion that pure science aims for the truth, while applied does not.

§3 Axiological Scientific Realism (ASR)

Nevertheless, within that project, many individual scientists doubtless had their own quite different conceptions of the criterion of success, and a political commentator might say that the criterion of success was defeat of the Japanese, or post-war supremacy over the Russians. So what was *the* criterion of success for the Manhattan project? There is surely more than one answer to this, depending upon perspective. Even if there is a written criterion of success somewhere it will probably not be unique – the criterion specified by the Pentagon in their funding documents was probably different from that to be found in Oppenheimer's project documents. Lest I be accused of picking on *applied* science, the same observations could be made concerning the LHC project.[15]

Let me consider the question of what exactly could be meant by the notion of there being a 'criterion of success' common to all scientific theories. What exactly is meant by this? Presumably it is intended to be a means of establishing whether or not a theory is successful, but in that case it is not clear that there will be one single such criterion from all perspectives – what counts as success for one scientist may seem like failure for another. We have already seen that what counts as success for a scientific project is perspective dependent, and it is reasonable to think that the same is true of scientific theories. It may be reasonable to ask who or what is supposed to have set this criterion, but in the case of chess that would be to miss the point, for the criterion of success for that game isn't set by some person, but is, rather, definitional of what chess *is*. So maybe the idea is that there is, similarly, a success criterion that is definitional of what science *is*.

---

[15] Is the LHC criterion of success confirmation/disconfirmation of the existence of the Higgs boson, the development of a 'grand unifying theory' via supersymmetry, or the development of commercial spin-off in the form of an expanded scientific infra-structure? In reality, aren't these all separate criteria of success?

§3 Axiological Scientific Realism (ASR)

However, as we have discussed, there is a big difference between chess and science – we can look up the rules of chess in a book, and that will tell us the criterion of success, but we cannot do that for science, nor for scientific theories.[16] So what is it to claim that there is a criterion of success for science – i.e. for all theories? We may again feel driven to ask scientists themselves. After all, although someone could discover the success criterion for chess in a rule book, an easier way would be to ask a chess player since, as Rosen points out, 'the aim that determines what counts as success is normally constituted by the conscious understandings of the participants.' (1994: p146). However, while this works for chess because there will be a common understanding among all participants due to chess having well-defined rules, I have argued that science is not like this, lacking both unity and clearly defined rules. So we cannot discover the criterion of success for science by asking scientists, even if that was a legitimate method of furthering a philosophical discussion, which it is not.

So one is forced to conclude that the criterion must be set by the philosopher who makes the claim. In other words, the philosopher who claims that the criterion of success for science is that it Φ is simply stating her vision of what science is, and she does this by stating a criterion of success which should be understood as that philosopher's interpretation of the nature of science. For example, if the scientific realist claims that the criterion of success for science is attaining true theories, then although stated as a claim concerning what *science* is, it is better understood as stating what it is to be a *scientific realist* – it is to endorse a policy of regarding a

---

[16] van Fraassen grants that "there is a strong disanalogy between chess whose rules and criteria for success are uncontroversially defined by official rulebooks and such large and vaguely circumscribed cultural phenomena as 'the game of science.'" (1994: p186)

scientific theory as successful if it is true. Similarly, to be a constructive empiricist is to endorse a policy of regarding a scientific theory as successful if it is empirically adequate. So one is forced to conclude that the axiological claim is, in effect, simply a declaration of the stance or attitude towards science taken by the philosopher who makes the claim, rather than a testable (or even arguable) claim about the objective nature of science or its theories. The philosopher who purports to say 'science is Φ' by way of 'science aims to Φ' ends up saying that she adopts a policy (stance, attitude, ...) as to what she will count as success in science, and that policy is independent of whatever policies individual scientists may adopt. This is not to rule out the possibility of reasoned debate between philosophers adopting different stances, but it suggests that the way to understand a philosopher's axiological claim is as a declaration of *their* interpretation and understanding of the scientific activity. This makes the axiological claim different in kind from claims such as RC1, RC2, RC3 since these latter are claims about the nature of scientific theories.

§3.4 van Fraassen on the Aims of Science

There is very little discussion in the literature as to what precisely is meant by axiological claims. The works of Musgrave, Lyons, or Sankey, give no indication of what they mean by a phrase of the kind 'science aims', nor any recognition that such a phrase might be problematic. Musgrave's teacher, Popper, says a little in his essay 'The Aim of Science' (1972: pp191-205) but it is unhelpful. The opening paragraph seems to completely undermine the stated aim of the essay before it even gets started. He begins:

§3 Axiological Scientific Realism (ASR)

> To speak of 'the aim' of scientific activity may perhaps sound a little naïve; for clearly, different scientists have different aims, and science itself (whatever that may mean) has no aims. (1972: p191)

Quite so indeed! Nevertheless he goes on to suggest that we do feel that there is 'something characteristic of scientific activity' (*ibid.*) and that:

> since scientific activity looks pretty much like a rational activity, and since a rational activity must have some aim, the attempt to describe the aim of science may not be entirely futile. (*ibid.*)

Well firstly, the claim 'a rational activity must have some aim' applies to an individual and not to a group activity – after all, he has just said himself that 'different scientists have different aims'. However, in addition, just as Fine pointed out earlier that it is a mistake to go from 'they all have aims' to 'there is an aim they all have', it is a logical blunder to go from each scientific activity having an aim, to the assumption that there is one aim shared by all scientific activities. After this inauspicious start the remainder of Popper's essay is devoted entirely to explicating the concept 'explanation' and says nothing about what might be meant by 'science aims'.

In contrast, van Fraassen has given some indication of what he means by 'science aims' and I have already referred to some of his remarks. There is also a subtle and detailed discussion by Gideon Rosen (1994), and an equally nuanced reply by van Fraassen (1994) which are the only sources in the literature that pay serious and detailed attention to these questions. This dialogue casts light on what van Fraassen means by claiming a criterion of success for science, so extended quotation is in order. A key remark of his is this attempt to differentiate between the aim of science and the aims of scientists. He claims that there is a distance between 'what the scientist pursues in his or her work and his motives or intentions in undertaking this work.' and he continues:

> some do it to discover the true laws of nature; many more do it to discover the structure of certain unobservable entities which they believe to exist. But the 'it' that they do, I claim, is work whose criterion of success in actual practice is empirical adequacy of the theories produced.
>
> These scientists with their very different motives and convictions participate in a common enterprise, defined by its own internal criteria of success, and this success is their common aim 'inside' this cluster of diverging personal aim. How else could they be said to be collaborating in a common enterprise? The question is only what that defining criterion of success is. (1994: p182)

He goes on to say that we cannot assume that 'what all the participants say they are doing is what they are doing.' and he then claims that others would agree with him on this if the question being considered was not science but art or religion (*ibid:* p186). The final quote I want to make is perhaps the most telling:

> Clausewitz' doctrine of war: [the aim of] war is the continuation of diplomacy by other means. This does imply: 'the soldier's aim, the criterion of his success, is the continuation of [his/her country's] diplomacy'.

§3 Axiological Scientific Realism (ASR)

But would Clausewitz have been refuted if all the generals canvassed insisted (in all evidence, sincerely) that their aim in war was to defend civilization, to cover oneself and one's country with glory, or to bring about universal peace and brotherhood, while their countries' diplomacy was clearly aimed at mercantile advantage and domination? (*ibid.*)

What are we to make of all this? To begin with, we can clearly see van Fraassen's conflation of the *aim* of X with what X *is* because he thinks that 'war is the continuation of diplomacy' is equivalent to 'the aim of war is the continuation of diplomacy'. So similarly he will think 'the aim of science is empirically adequate theories' is equivalent to 'science *is* empirically adequate theories'. However, I wonder if the question 'what *is* science?', with its essentialist overtones, is the right question for an empiricist, and perhaps it is not the best way of giving an interpretation of science. Faced with any complex social phenomenon, asking what it *is* isn't the best way to derive an interpretation and understanding of that phenomenon. Asking what it *does* is likely to be more successful. It might be objected that there is no gap between the questions 'What *is* X?' and 'What does X *do*?'. However, if one considers, for example, another complex social entity – the EEC – it seems to me that the questions 'What is the EEC?' and 'What does the EEC do?' would not yield at all the same kinds of answer.

van Fraassen seems to insist that there is some unique intentional attitude – an aim – that defines an enterprise, but which is independent of the intentional attitudes of the people who take part in that enterprise; and, moreover, that this unique intentional attitude is definitional of what that enterprise *is*.

§3 Axiological Scientific Realism (ASR)

I have already suggested that it is a mistake to think that there is one unique way of describing any enterprise, and I think this applies to van Fraassen's use of the Clausewitz doctrine. I do not think there is one single aim that uniquely describes any specific war, but even if there was, we have no reason to assume that such an aim would be common to all wars. I think it is a mistake made by both Clausewitz and van Fraassen to think otherwise. Clausewitz's doctrine is an interesting and illuminating view of the nature of war from the particular perspective of the diplomat. It rings true in many ways, but false in others, and is merely one way of describing war, and cannot be regarded as the unique statement of what war *is*, for the term 'war' is surely another example of family resemblance. I would claim that there is no unique way of saying what one specific war *is*, so there certainly cannot be an essentialist statement *tout court* of what war in general *is*. Of course, the reason van Fraassen produces this example is because it justifies the claim he wishes to make – that scientists *really are* aiming for empirical adequacy even if they say they are aiming for truth.

I can summarise my opposition to general criterion of success claims for science in terms of three main objections: First, of course scientific projects have aims *qua* criteria of success, but they cannot necessarily be defined by one single such aim that is independent of perspective. Second, it is a mistake to see all scientific projects as belonging to a unified science having an essence the same as the essence of all those separate projects. Science is a multifarious and diverse activity lacking unity other than in the sense of satisfying the practical needs of various people in various situations – needs concerned with being able to predict phenomena and make pragmatically useful interventions. Thirdly, the attribution to a project of an intentional attitude in the form of a criterion of success seems dubious, but to then also say that such an intentional attitude is independent of, and quite different from, those of the project's participants seems even more so. For these and other reasons I have discussed, I question the usefulness of claims concerning a criterion of success for science. Given that I have already ruled out the teleological construal of 'aims', that rejection extends to the whole idea of attributing an aim to something called 'science'. It is reasonable to speak of the aim(s) of an individual scientific project, though even then such aims may depend on perspective. However, that does not legitimate the attribution to science of aims, let alone one single unitary aim.

§3.5 Objections

§3.5.1 Can Science be Aimless?

This may all seem counter-intuitive – someone might say that if science does not have an aim then it must be aimless, but that does not follow. It is obvious, and I have conceded, that scientific projects have aims and criteria of success, but it does not follow that *science* has *an* aim. It may well be correct to claim a degree of commonality among the various practices and methods comprising the totality of science, or that the results produced by the sciences show some commonality. That is what claims such as RC1, RC2, RC3 amount to, but such commonality of practice and result is not best described in terms of aims and criteria of success.

§3.5.2 Common Method Entails Common Aims

Another objection might be that all the sciences use common methods, and that commonality of method entails common aims qua criteria of success. I reject this argument for the following reasons:

A group of artists may display common method in their use of easel, canvas, brushes, paints, as well as similarity in the way the paint is applied or the subject is studied, but no one would suggest that such artists share a common aim or criterion of success. Similarly, all philosophy PhD students would use similar methods, such as discussing with their supervisors, pursuing bibliographic references, presenting papers to their peers, attending conferences, but their theses do not share a common criterion of success. The similarity of their method would only entail a common criterion of success if their adherence to these methods is, *of itself*, the measure of success, but it is not. The criterion of success for a PhD thesis is not the student's adherence to conventionally agreed methods. Such adherence may ensure that a PhD thesis attains a satisfactory level of academic excellence, but that is not the only criterion of success, perhaps not even the most important one. In both of these cases, be it painting or thesis writing, whether or not the outcome is successful is beyond simple definition and could not be summarised in the form of a single criterion of success related to the methods used.

Similarly, in the case of the sciences, the only common criterion of success that might be entailed by commonality of method would be that pertaining to whether or not a project has adhered to standard approved methods – i.e. something of the form '*the* criterion of success for a scientific project is that it conforms to approved scientific method', and that is obviously false. In §9.3 I argue that scientific method is conducive to the production of theories with good predictive ability. If method is common across all the sciences then one could reasonably expect that theories with good predictive ability will also be common. However, while that might well suggest that predictive ability is *a* criterion of success in all the sciences, it would not entail that it is *the* criterion. For if it was *the* criterion, then, faced with a choice between two theories, the one with the greater predictive success would always be chosen regardless of any other consideration, but it is entirely plausible that a predictively successful but clumsy and inelegant theory might be rejected in favour of another

§3 Axiological Scientific Realism (ASR)

theory that is marginally less predictively successful but displays more of what van Fraassen calls the 'pragmatic virtues'.[17] Consequently, predictive ability could only be *one* criterion of success among others. Moreover, any proposal that predictive success is *the* criterion would hardly be conducive to the realist cause! Just as with painting and PhD theses, even though there may be adherence to common method, leading to certain common outcomes, it does not follow that there is a unitary aim *qua* criterion of success. This illustrates the way in which reference to aims for science is unhelpful and that a description and interpretation of its results is more productive.

So the argument that common scientific method entails common aims seems weak. Realists who believe that the criterion of success for science is truth may reply that the method of science itself is intimately concerned with the attainment of truth. However, such a reply will not get them very far because even if that is conceded, as I show in §3.7.4, it would only be truth concerning the observed phenomena. As I show there, realists fail to demonstrate any connection between method and theoretical truth. Consequently, all the realist could claim is that the methods of science are intimately concerned with the attainment of predictive ability, thus lending *prima facie* credibility to some anti-realist axiological claims, but not their own.

---

[17] Predictive ability was not the criterion of success that led to the triumph of the Copernican system over the Ptolemaic, but simplicity and what Kuhn called the astronomer's 'aesthetic sense' (1957: p181).

§3.6 Conclusions: The General Axiological Claim

What is the motivation that leads these philosophers – realist and anti-realist – to speak in terms of the aims of science? van Fraassen alone suggests a reply to this, namely that they want to understand what science *is*, and they then translate that into a wish to understand what science *aims at* on the assumption that what an enterprise *is* can be determined by understanding what it aims *at*. We may contrast this desire to establish what science *is* with the approach taken by the COSR theorist who wishes to establish something about scientific theories. As suggested earlier, posing the question 'what is science?' may be the wrong way to begin, and only leads to these problematic issues of aims and success criteria.

The task of any 'philosophy of X', regardless of whether X is science, religion, history, or anthropology, is to provide an alternative interpretation of those activities and of what people within those activities say. van Fraassen should agree, since he says that 'a philosophy of X proposes an interpretation of X' (1994: p190). Regardless of whether one is realist or anti-realist, this discussion of constructive empiricism by Bradley Monton & Chad Mohler (2008: §1.4) seems to capture admirably what it is to give an interpretation of science:

> Like the interpretation of any human activity, constructive empiricism is constrained by the "text" of the scientific activity it interprets. Within those constraints, it succeeds or fails according to its ability to provide an interpretation of science that contributes to our understanding of science, making intelligible to us various elements of its practice.

I submit that such an interpretation of science is best done without unhelpful reference to aims, or criteria of success. An

§3 Axiological Scientific Realism (ASR)

interpretation of science is best given by offering an alternative way of understanding what its practitioners are doing and what its results give us. In terms of scientific theories, this translates into offering an alternative understanding of what those theories actually achieve and how they are best understood. Thus, instead of proposing interpretations of the form 'science aims to develop true theories', or 'science aims to develop empirically adequate theories', interpretations of the following form would be clearer in their meaning: 'science develops theories which are best understood as being true', or 'science develops theories which are best understood as being empirically adequate'. While these could be understood as answers to the question 'What *is* science?', they are perhaps best understood as being concerned with what science *does*, and the latter yields answers that are more readily tested and debated than the former. In this respect, the COSR theorist seems to at least start out on a more secure foundation.

Nevertheless, a charitable interpretation of the general axiological claim is available, and we have already seen what it is. I suggest that axiological claims can best be understood as placing those philosophers who make them 'in a position to make sense of those activities which we all agree are part of science.' (van Fraassen, 1994: p190) Dubious talk of 'aims' can then be put aside and we can take these philosophers to be offering their interpretation and understanding of what science results in, whether realist or anti-realist.

I have attempted to show that close scrutiny of scientific axiological claims suggests they are not what they appear to be and that axiological claims are not a helpful way to characterise science. I will now further criticise the specific realist claims that are, after all, my target. My intent is to show that realist axiological claims are not just unhelpful, but misguided.

§3.7 The Axiological Formulation of Scientific Realism
§3.7.1 Introduction

Let me first dispose of the possibility of construing the axiological claim as normative. That could not be excluded in the earlier more general discussion, but in the specifically realist context, it is clear from the words of Musgrave (1985, 1988, 1996, 2006), Lyons (2002, 2003, 2005, 2006), and Sankey (2008) that a normative construal is not intended. To give just one example, Lyons (2005: p167) claims that his axiological postulate is '… the richest account of the scientific enterprise', but a normative claim couldn't be described as an 'account of the scientific enterprise'. Anyone who reads these works will see very clearly that these philosophers are not proposing normative claims. Moreover, the normative construal can be discounted in the same way that Rosen discounts that construal of van Fraassen's constructive empiricist claim. Both scientific realism and constructive empiricism are philosophical positions, addressed to philosophers of science, not to scientists. If the claim was normative then we might expect to find these philosophers interested 'in changing the minds of the agents whose attitudes constitute the aim of science, namely the scientists themselves.' (Rosen, 1994: p148) Yet we do not find evidence of either Musgrave or Lyons wishing to influence the intentional attitudes of scientists,[18] and van Fraassen is explicitly content for them to be realists.

I now recall the axiological claims of scientific realists, and at this stage in the discussion I am concerned only with what realists claim, so I set aside van Fraassen's views, while retaining his criterion of success notion:

---

[18] Though this is not true of Popper.

§3 Axiological Scientific Realism (ASR)

Musgrave: The aim of a scientific inquiry is to discover the truth about the matter inquired into.

Lyons: Science seeks true theories and is justified in doing so.[19]

Sankey: The aim of science is to discover the truth about the world.

I regard the teleological construal of ASR as having been ruled out, and the criterion of success construal, AC4, can be re-stated in explicitly realist form:

ACR4: The criterion of success for science is that it
discovers the truth (Musgrave)
discovers the truth about the world (Sankey)
develops true theories (Lyons)

In addition to the negative critique of axiological claims in general, there are additional decisive objections to ASR.

§3.7.2 Truth About What? Observable/Unobservable Again

Let me recall Musgrave's claim:

MC: the aim of a scientific inquiry is to discover the truth about the matter inquired into

This risks triviality, for we may wonder whether all enquiries, by definition of the word 'enquiry', set out to 'discover the truth about the matter'. What would an enquiry be like that did not set out to discover the truth about the matter enquired

---

[19] As noted on p55n2, Lyons moves from truth as the aim of science to a more complex formulation. For that reason I will make little reference to his views from this point on.

§3 Axiological Scientific Realism (ASR)

into? That sense of triviality arises from a lack of clarity as to what is meant by 'the truth about the matter'. ACR4 inherits the same problem of ambiguity. For what might Musgrave mean by 'the truth', or Sankey by 'the truth about the world'? Both of these are too vague to convey much. What exactly does Musgrave intend by the phrase 'discover the truth'? There is one sense in which 'truth' is ubiquitous and applicable to almost everything, but in that sense it is trivial. If snow is white then of course it is true that snow is white; if quarks exist then it is true that quarks exist; but if this is all that is meant by the 'truth' that science is supposed to aim for, the claim doesn't amount to much as the following example of a 'scientific inquiry' shows:

In 1590 Galileo went up the tower of Pisa to conduct an enquiry, and the matter enquired into was whether he was right to believe that objects of different weights would reach the ground in the same time. Is it useful to view such an inquiry as concerned with 'truth', at least in anything other than this trivial sense? Isn't it more reasonable to think of Musgrave's 'discovers the truth' here as meaning 'establish whether objects of different weights really do reach the ground in the same time'? If they do then it will be *true* that objects of different weights reach the ground in the same time, but that's back to the triviality of 'truth'. Galileo's aim was not truth, but to discover whether objects of different weights would reach the ground in the same time. Claiming that he must have therefore sought the truth about this seems like a device intended to give the appearance that many scientific projects with widely differing aims are united under the unitary aim of truth.

Most realist philosophers esteem science as the epitome of knowledge acquisition, and the latter has been equated with the search for truth by philosophers since Socrates. Consequently, those philosophers are inclined to read MC as saying that the aim of science is the search for truth.[20] However, we don't have to read MC in that way since 'discovers the truth about the matter inquired into' could equally be taken to mean 'get to the bottom of the matter inquired into' or 'discover how things behave in the matter inquired into' – both of these would suit the Galileo case nicely. I suggest that MC can be just as well interpreted as meaning that science aims to discover how things actually do behave – which will of course lead to accurate predictions and the likelihood of useful interventions. Does Galileo seek 'the truth' or does he seek to be able to accurately predict what will happen in various circumstances? Whatever may inspire Musgrave's way of speaking, it need not actually involve any 'truth' other than in a trivial sense.

In a case like that of Galileo's weights, bringing 'truth' into the picture seems to add nothing. However, there is something about the Galileo case that is different from most scientific inquiries, for most scientific theories, as well as making statements as to observable behaviour, as Galileo does, also include theoretical terms that can be taken as referring to putatively existent unobservables. Now if a theory deals only in observable behaviour it is fairly straightforward to say that there is an isomorphism between some statement of the theory and some aspect of the world. Consider, for example, the 'theory' describing the distance dropped by Galileo's weights, expressed by the simple equation:

$$d = \tfrac{1}{2} g\, t^2$$

Here d represents the distance dropped after time t, and g is the earth's gravitational constant. This is a mathematical

---

[20] van Fraassen comments on philosophy's tendency to project its own self-image: 'This is but part of the mistaken but very common projection of philosophy of its own enterprise into science, art, religion, and everything else it studies.' (1994: p190) This seems particularly apposite concerning science.

equation, the two terms of which can be taken to be isomorphic with the actual distance dropped after an actual time. So this theory is both descriptive – enables predictions – and could also be said to be *true* in a correspondence sense, the 'correspondence' being expressed by this isomorphism. In addition, the correspondence could be verified by the simple step of taking measurements. We have a model, the equation, and observable entities in the world – distance and time, and in a sense the correspondence itself can be observed. There is no mystery here, and its clear why 'truth' adds nothing to the discussion. However, if the theory includes theoretical terms that refer to entities whose existence is postulated then declaring the theory 'true' becomes problematic. For whilst 'it is true that snow is white' reduces to 'snow is white' and an isomorphism can be demonstrated between 'snow is white' and the world, this does not work for 'it is true that quarks exist'. Whilst the deflation move can be made – to 'quarks exist', establishing an isomorphism between this and the world cannot be done by any method independent of the theory concerned.[21]

Returning to the ambiguity of both Musgrave's and Sankey's claims, about what, precisely, is the truth sought? Presumably it is agreed by all that science has to 'capture the phenomena', which entails that science must attain/discover the truth about the observable. So their claim would be trivial if it concerned only the observable. I suggest that both Musgrave and Sankey avoid saying what they really intend – that a scientific enquiry aims to discover the truth about the unobservable aspects of 'the matter' or 'the world' as well as the observable, which latter is hardly worth stating. Only when that is realised does the sense of triviality disappear. If we turn to the formulation

---

[21] This claim, and the more general issues concerning correspondence truth, are discussed more fully in §8.5. Note also that both Musgrave and Sankey are explicit in their allegiance to the correspondence theory of truth.

that Lyons starts out with – 'science seeks true theories' – one again may wonder if this is a little trite. This triteness is again removed when it is realised that it means something like 'science seeks theories which are true in their statements about the unobservable as well as the observable'. For it is surely obvious that a scientific theory needs to be true concerning the observed phenomena.[22] This suggests that the apparent avoidance of any reference to the observable/unobservable distinction by axiological formulations of scientific realism (see p37n13) can be questioned and that, in reality, this form of scientific realism is just as dependent upon that distinction as are the others.

§3.7.3 Is Truth as Sole Aim of Science Credible?

The next issue I want to raise concerns the sense of incredulity one may experience if one takes ACR4 literally (see p 79). For it seems to say that there is just one unitary criterion of success throughout the whole of science, in all its activities, namely the attainment of truth (for both observable and unobservable). This proposal seems to immediately negate the suggestion, previously mentioned, that scientific projects frequently aim to make successful predictions of phenomena with a view to enabling interventions in the natural world. Many would think it is simply obvious that large numbers of scientific projects have precisely that aim, and some, myself included, would think this is the engine that has driven the growth and success of all the sciences. Moreover, all sides to the debate agree that science is successful; yet that success is exemplified by its successful predictions. However, we are told that its sole aim was not those successful predictions, but truth. This is surely deeply implausible, a point I shall return to later. We are told that

---

[22] Perhaps it seems obvious because we have an intuition that the prime concern of the sciences is to enable accurate predictions, not the attainment of truth.

§3 Axiological Scientific Realism (ASR)

successful prediction isn't really an *aim* of science at all, for the claim is not that attainment of truth is *one* of the aims of science, or even the most important, but that it is *the* aim of science. The implication is that all other apparent aims are either illusory or entirely subservient to the aim of truth. At the very least the prospect of a clash between philosophy and sociology looms. For suppose the sociologist of science suggests that Newton's theories were derived when they were derived because of mercantile pressures, which in turn suggests aims other than truth. According to ASR this can be ruled out, for the aim of science is always and only the attainment of truth. We might also recall the Manhattan project again, where we are told that, regardless of what we thought, in fact its aim was the attainment of truth; and if the ASR theorist seeks to rule out such projects on the grounds that they are not 'pure science' she will once again appear to be question-begging as her grounds for saying this will doubtless be that such projects do not aim for the truth.

Moreover, what of the other conventionally assumed objectives that govern scientific method? I refer to the scientists' requirements of a theory – economy, elegance, explanatory power, unifying power, etc. Would a clumsy, inelegant, and difficult to use theory that is believed to be 'true' be regarded as successful by most scientists?

Although Laudan agrees with the idea of science having aims, he is a pluralist and argues that scientists' multiple aims vary with history.[23] Sankey (2008: p96) concedes aim pluralism to Laudan, saying that 'a multiplicity of aims may be pursued by scientists.' However he then qualifies this by saying that the 'various other cognitive aims which may be pursued by

---

[23] As examples: infallible knowledge, high probability, elegance, and Newton's aim of revealing divine agency at work in the world (1984b: p51ff, 1996: p129).

scientists may be understood as subordinate aims which subserve the overriding realist aim of truth.' (*ibid.*)

This is no real concession since it makes all other possible aims merely instrumental to the achieving of truth – the scientist aims for elegant theories only because he believes that elegant theories are more likely to be true. I repeat the question – would a clumsy, inelegant, difficult to use theory that is believed 'true' be regarded as successful by most scientists? According to Sankey it would have to be, because the aims of elegance and ease of use merely subserve the aim of truth. Again we encounter a large divergence between the intentional attitude attributed by a philosopher to science and the intentional attitudes of actual scientists. In addition, one may wonder what reason Sankey might have for thinking that theory elegance and ease-of-use will in fact subserve the aim of truth, which leads on to my next objection.

§3.7.4 What Connects Methodology and Truth?

I turn now to the question of scientific methodology. I discuss this in detail in §9.3 where I argue that the success of science is due to its methodology, so there is a direct connection between that methodology and the ability to make successful predictions. How would the realist propose to establish a connection between her proposed aim of truth and the methodology that science actually employs? It is an empirically observable fact that the methodology of all the sciences involves procedures governing, for example, experimental practice and theory appraisal. A theory is constructed that will predict some phenomena and work can be done to verify those predictions. If incorrect the theory is either modified or discarded. If correct then progress has been made and work often begins aimed at extending the theory. All this takes place in a public arena in which results are appraised through a rigorously applied peer review process. This is a greatly simplified approximation to the complexity

§3 Axiological Scientific Realism (ASR)

and diversity of actual scientific method, but it captures the obvious connection between method and truth concerning observed phenomena. A successful theory that has been subject to scientific methodology just has to be true of the observed phenomena, at least in most respects. Consequently, a scientist who complies with these agreed methodological procedures in producing a theory has warrant for accepting that theory and believing it to be (largely) correct regarding the observed phenomena.[24]

Now I have been critical of the very notion of interpreting the activity of the sciences in terms of aims, but on the basis of empirical examination of scientific practice, if forced to speak of an aim for science, it would surely be that of producing theories that are true concerning the observable phenomena, for actual scientific methodology at least lends that some credibility. Moreover, I could then say that the methodology used subserves *that* aim – of truth concerning the observed phenomena – which is reminiscent of van Fraassen's *constructive empiricism*.[25] For the reasons I have given, I reject this way of speaking in terms of aims, but I concede the link between method and predictive success. Now the problem for the axiological realist becomes apparent, for she claims that science aims for the truth regarding observable *and* unobservable. So in addition to all the problems I have raised for axiological claims in general, the axiological realist needs to show how scientific methodology is connected with theoretical truth – truth regarding the unobservable. Moreover, the connection will need to be strong, for whilst it

---

[24] I show in §9.3 that those practices and procedures are conducive to the production of theories that capture the phenomena – theories that are true concerning the observable.

[25] Though not the same – truth about the observ*able* goes beyond truth about the observed, so despite van Fraassen's complaint that realists take too much risk, he also takes risks. Strictly, his empiricism only entitles him to truth regarding the *observed* phenomena.

§3 Axiological Scientific Realism (ASR)

is clearly reasonable to say that scientific methodology does literally subserve truth regarding the observable, it is hard to see how that could be said of truth regarding the unobservable. For even if all scientists on a project professed themselves to be aiming for truth concerning the unobservable, it would not be obvious that the methods they employ are *necessarily* conducive to obtaining that truth. As Sankey himself points out

> The truth of the non-observational content of theories transcends empirical verification, hence cannot be established by direct observational means. (2008: p118)

For the realist (certainly for both Musgrave and Sankey), truth involves correspondence between theory and world – both observable and unobservable, and that may exist regardless of whether methodology has warranted belief in it. So what is there about scientific practice and methodology that warrants belief in the attainment of truth concerning the unobservable?

Sankey (2008: pp125-143) sees this as a problem of central importance for scientific realism, and discusses it in the specific context of Musgrave's brand of realism. Sankey concedes that there is no obvious link between scientific method and truth and that the scientific realist must demonstrate one, and he examines two alternative approaches that have tried to construct such a link. The first of these he describes by way of Lakatos's critique of Popper, where he called for 'a whiff of inductivism' which, contra Popper's anti-inductivism, would enable a high degree of theory corroboration to be taken as entailing a high degree of verisimilitude – in other words, correct method entails truth (Schilpp, 1974: pp254-260). This of course is the No Miracles

Argument (NMA) in disguise – we should infer truth from success.[26] Sankey suggests that the inductivist nature of this proposal may be why both Popper and Musgrave do not take this option. The second way in which a link has been established between method and truth is the 'internal realism' of Putnam and Ellis which, in effect, makes the link analytic by making truth epistemic and theory-relative. Obviously realists such as Musgrave and Sankey could never accept this proposal, as they embrace a non-epistemic account of truth – correspondence with a mind-independent world. Musgrave, for example, has argued that internal realism amounts to idealism (1996). In addition to these two alternatives, we may add a third from an earlier Sankey paper in which we are asked to suppose that some theory

> satisfies a broad range of rules of method to an extraordinarily high degree ... It unifies previously disparate domains, ... maximizes simplicity and coherence. ... the best explanation of such success is that the theory provides an approximately true description of the way the world is. (2002: §9)

This amounts to yet another NMA variant. After some discussion, Sankey (2008: pp130-137) concludes that Musgrave fails to establish the link between method and truth, and Sankey then professes the belief that the only way to successfully establish the link from method to truth involves making some metaphysical assumptions. He gives two examples – the first is Rescher's *methodological pragmatism*, which Sankey (2008: p139) describes as claiming that a pragmatically successful method that is systematically erroneous can be rejected as being incredible. As Sankey says

---

[26] Discussed in detail in §8.

(*ibid:* p138), 'Rescher takes certification by rules of method to warrant acceptance as true'. Rescher's reasons for this claim seem to boil down to another variant of the NMA.

The second example given by Sankey is Kornblith's *realist metaphysics of natural kinds* in which the success of science is taken to show that natural kinds are the ground of induction (2008: p140). It isn't clear what Sankey thinks he achieves by a move like this, that could only ever convince himself and other realists. For no anti-realist is likely to accept either of these proposals – one is based upon a metaphysics which anti-realists are unlikely to accept, and the others are all variants of the NMA, which again no anti-realist would accept. Moreover, Musgrave himself probably wouldn't accept this as a way of bolstering his arguments for axiological realism. In effect, Sankey replies to an argument against axiological scientific realism by employing arguments that no anti-realist would accept, and only *some* scientific realists would accept.

What can we deduce from this? Sankey's investigation of Musgrave's claims regarding the link between method and truth show that Sankey (a realist in terms of both ASR and COSR) finds it impossible to establish a link from method to truth without employing metaphysical principles that would be unacceptable to anti-realists and to Musgrave himself. Now the ASR claim is that the aim of science is truth – regarding both observable and unobservable – and I can reasonably claim that, without giving anti-realists reason to believe there is some link between method and truth, the ASR claim appears deeply implausible. No such reason has been given.

We shall come to see this theme throughout this essay. For the whole point of the pessimistic induction, as I present it, is to establish the absence of any link between predictive success and approximate truth (see §5.4) but predictive success is precisely what is conferred by proper conformance to established scientific methodology.

§3.7.5 ASR Requires COSR

The arguments above aim to reduce the credibility of the realist axiological claim. In this final section I wish to argue that the claim of ASR theorists, Musgrave and Lyons, to reject COSR cannot be upheld since it leads to an inconsistent position that can be avoided only if they also adopt the COSR view.

ASR may seem a rather weak claim, immune to criticism; for it makes no claims concerning the truth of any scientific theories, and doesn't really seem to claim very much for science.[27] According to ASR science may never have produced any true theories even though it always aimed to do so, which would hardly suggest a successful enterprise. Nevertheless, as we saw earlier, both Musgrave and Lyons claim to oppose views such as COSR, which claims that most scientific theories are true/approximately true. In addition, it is easy to see how to make arguments both for and against COSR, for example by investigating the historical record, but no such investigation can touch ASR. Indeed, since investigation of the intentional attitudes of scientists is also excluded, there doesn't seem to be any way of either supporting the claim or refuting it, a point made already concerning van Fraassen's axiological claim.

As I said above, we can assume that all sides of the debate think science is successful. If we now consider the axiological claim in its criterion of success form, then van Fraassen would regard a scientific theory as successful if it is empirically adequate, so it follows that his interpretation of science entails that it yields theories that are, in general, empirically adequate. In other words, his axiological claim reduces to a

---

[27] As we have seen, it says more about the beliefs of its holders than it does about science.

§3 Axiological Scientific Realism (ASR)

claim about scientific theories. He might disagree with such a reduction, but it would not be contradictory for him, for it simply amounts to saying that, at a given time, successful scientific theories are empirically adequate, with which he would readily agree.

By similar logic, the philosopher who espouses ASR regards scientific theories as successful if they are true, from which it follows that she must think scientific theories are, in general, true (or approximately so). For how could someone believe that the criterion of success for theories is truth, *and* that science is generally successful, and yet not believe that our current successful theories are generally true? This is too quick because there may be two different senses of the word 'success' in use here. The belief that science is successful could amount to either of the following claims:

S1a: science is successful in that its theories are true

S1b: science is successful in that it makes correct predictions

In addition, we have the criterion of success claim:

S2: the criterion of success for scientific theories is truth/approximate truth

If the ASR theorist claims adherence to S1a then she must think that most scientific theories are true, and thus she should agree with COSR, which would of course be fully coherent with S2. Now in fact S1b seems the more plausible construal and it would be more charitable to assume this is what the ASR theorist believes, as this is surely how the claim that science is successful is normally understood (and is the construal made throughout this essay, see p28n5). However, there is then a rift between her criterion of success for science in total (correct predictions) and her criterion of success for

§3 Axiological Scientific Realism (ASR)

specific theories (truth/approximate truth). Unless there is a strong argument for a link between those two – correct predictions and truth/approximate truth – that position seems inconsistent as it depicts science as constantly aiming for truth, but achieving correct predictions. The only way out of that for the ASR theorist is to assume that there is some link between correct predictions and truth, that achieving correct predictions *entails* achieving truth/approximate truth. However, such a link – from empirical success to approximate truth – is precisely what, as we shall see later (§5.6), the PI argument succeeds in refuting – approximate truth does not entail predictive success, neither does predictive success entail approximate truth. I have shown previously that the link between scientific methodology and empirical adequacy is clear to be seen, so again the question of a link between that methodology and theoretical truth is the problem confronting scientific realism.

Even if such a link *could* be established, and it has not been, it would only result in forcing the ASR theorist to adopt the COSR position of believing that most scientific theories are true. However, we have seen that ASR theorists claim to reject the COSR option, but those claims cannot logically be upheld as they would place them in the position of taking wholly different attitudes to the success of theories and the overall success of science. Consequently, despite what they say, their belief in ASR needs augmenting by belief in COSR.

The conclusion here is that ASR theorists must either embrace COSR or admit to an unacceptable gap between the criterion of success they advocate for theories and that which they accept for science as a whole.

§3.8 Conclusion

I have argued that axiological claims in general are unhelpful. The axiological question 'What is the aim of science?' must be interpreted as meaning 'What is the criterion of success for science?', and is based on the question 'What is science?' This assumes more commonality among the various sciences than is the case. In addition, it suggests an essentialist view of science rather than a more appropriate family resemblance conception of *the sciences*. Moreover, once we come to understand that what philosophers are trying to achieve with their axiological claims is an interpretation and understanding of the activities of the sciences, it becomes clear that this is best achieved by claims of the semantic, metaphysical, and epistemic kind, such as RC1, RC2, RC3.

Furthermore, if axiological claims in general are unhelpful, the particular axiological claims of scientific realists lack credibility for the several reasons I have enumerated here, most important of which is my argument that they must either accept a radical split between their criteria of success for theories and for science as a whole, or accept that they have to embrace that aspect of COSR which says that the majority of current scientific theories are true.

Holding the ASR position but not the COSR position is not a viable option.[28] Consequently, I take the refuting of COSR as, in effect, refuting ASR, and the rest of this essay is almost entirely concerned with COSR. Henceforward, although keeping in mind the distinction between COSR and ASR, I frequently refer to just 'realism', which can be taken as referring to COSR.

---

[28] We have already seen that both Musgrave and Sankey are, in effect, committed to COSR (p55n4 and p54n1 respectively).

§4 Verisimilitude, Approximate Truth, and Truth-likeness[1]

§4.1 Introduction

As a preface to the detailed discussion of arguments concerning Convergent Ontological Scientific Realism (COSR) it is necessary to examine the concept of approximate truth, and its cognates, verisimilitude and truth-likeness. This is because COSR is highly reliant upon this concept,[2] and arguments in favour of COSR cannot be discussed without immediately encountering it.[3] Let me recall a central claim of COSR as I have characterised it:

RC2 Scientific theories are typically approximately true and more recent theories are closer to the truth than older theories in the same domain. (See p44)

We shall see later that Laudan's Pessimistic Induction argument questions the vagueness of the phrase 'approximate truth' and Laudan criticises the fact that the realist 'needs more than a promissory note that somehow, someday,

---

[1] In this chapter I italicise the single word *truth* to differentiate it from approximate truth.
[2] Those who explicitly include such a notion in their formulations of realism include Boyd (1973, 1984), Chakravartty (2007, 2008), Ladyman (2007), Niiniluoto (1999), Psillos (2000), Putnam (1975c), Smart (1968).
[3] It could reasonably be claimed that ASR is not dependent upon the concept of approximate truth. As I mention below (§4.5) Musgrave (2006) is explicit in rejecting the concept. Lyons (2005: p167) claims that his axiological realism is independent of any epistemic claims involving approximate truth. Sankey's position is less clear, but that is understandable given his adherence to both COSR *and* ASR (see p51n1).

§4 Verisimilitude, Approximate Truth, and Truth-likeness

someone will show that approximately true theories must be successful theories.' (1981: p32)

In this chapter I shall make a more detailed criticism of the concept of approximate truth as it applies to scientific theories, and thus question the legitimacy of the central role given to it by realists. It is perhaps unsurprising that *truth* is of such importance for scientific realism, so we might expect to find *truth* simpliciter occupying centre stage. However, most realists accept that we are generally unable to assert that a given theory is true in every respect and without qualification, this being particularly so for theories having theoretical content. Moreover, there are the issues of idealisation (see glossary), experimental error, and so on. These factors all push the realist towards advocating something like truth-likeness or approximate truth in place of the *truth* simpliciter which is beyond her reach.

Let's also remember that we are not here talking about approximate truth applied to some single proposition such as 'there are 20 people in this room'. Such a case is relatively straightforward due to the fact that it is a single proposition and concerns easily observed, measurable facts. A scientific theory is very much more complex on two counts – firstly it includes a whole variety of 'propositions' all highly interrelated, with many expressed in mathematical form, and second because it includes theoretical postulates the *truth* of which can be known only via the theory – there is no god-like position we can adopt in order to see if they are true, let alone approximately true. What is it, then, to say that an entire theory is approximately true? I leave that question hanging for now.

Three recent book length defences of scientific realism devote an entire chapter or major section to the notion of

§4 Verisimilitude, Approximate Truth, and Truth-likeness

approximate truth.[4] Other advocates have used concepts like 'relative truth' (Leplin, 1984b: p215) or 'partial knowledge' (Trigg, 1980: p196) and these are best understood in terms of approximate truth. It is also worth noting that whilst the concept is of central importance for the realist, it has no intrinsic importance for the anti-realist who either denies, or withholds assent from, the *truth* of theories, let alone their approximate truth.

It is beyond the scope of this work to perform an analysis of all the many attempts to explicate the concept, Niiniluoto's comprehensive survey (1998) itself running to thirty pages. I shall briefly mention the central attempts that have been made at a formal definition of the concept of approximate truth, together with their shortcomings. However, my purpose will be to show how fifty years of work on these formal proposals has yielded no success and that even realists themselves concede the absence of any usable result. I will then examine and criticise two very recent attempts at an intuitive analysis of the concept. I will conclude that no notion of approximate truth exists that legitimates its key position in the arguments for scientific realism. From this it follows that the realist claim RC2 not only lacks justification, but also lacks determinate meaning.

§4.2 Formal Proposals

Niiniluoto (1998) views the history of research on this topic as divisible into three periods. The first began with Popper's verisimilitude proposals (1963: pp427-35) in which he was the first to attempt a formal definition. He proposed that successive scientific theories within some domain of

---

[4] Psillos (1999: ch.11), Chakravartty (2007: ch.8), Niiniluoto (1999: pp188-196).

§4 Verisimilitude, Approximate Truth, and Truth-likeness

investigation can show increasing levels of verisimilitude, and this can be expressed as follows: Suppose there is a temporally-ordered sequence of theories:

$$Th_1, Th_2, Th_3, \ldots Th_n, \ldots$$

For the theory Th_n, let the set of all its true observational consequences be Th_n^T and the set of all such false consequences be Th_n^F. Popper proposes that a comparative ranking of the verisimilitude of two theories can be given by comparing both their true and false consequences. He (1963: p233) claims that, assuming that the truth-content and the falsity-content of two theories Th_m and Th_n (m < n) are comparable, we can say that Th_n is more closely similar to the *truth*, or corresponds better to the facts, than Th_m, if and only if either:

(a) the truth-content but not the falsity-content of Th_n exceeds that of Th_m

or

(b) the falsity-content of Th_m, but not its truth-content, exceeds that of Th_n.

Or, in formal form: Th_n has higher verisimilitude than Th_m just in case either of the following statements is true:

a) $\quad Th_m^T \subset Th_n^T \text{ and } Th_n^F \subseteq Th_m^F$
b) $\quad Th_m^T \subseteq Th_n^T \text{ and } Th_n^F \subset Th_m^F$

Here $\subseteq$ represents set theoretic inclusion, and $\subset$ represents proper inclusion.

§4 Verisimilitude, Approximate Truth, and Truth-likeness

Though intuitively attractive, this was independently shown to be fatally flawed by two of Popper's colleagues – David Miller (1974) and Pavel Tichý (1974). I will not discuss their proof here, but they showed that in both (a) and (b) the two conjuncts cannot both be satisfied. They proved that, given these definitions, Th_n can have greater verisimilitude than Th_m only if Th_n is *true* simpliciter. In other words, any false theory cannot have more verisimilitude than any other false one.

However, there is another reason why Popper's theory could be of no use to realism even if it was not flawed. This is because he defined verisimilitude in terms of observational consequences, so the most his theory could show is that successive theories entail more true, or less false, observational consequences. But that would not help realism for it needs a concept of truth-likeness which includes the theoretical content of theories, not just their observational consequences.

After the failure of Popper's attempt to define verisimilitude, two other families of accounts of approximate truth emerged – the 'possible worlds', and 'type hierarchy' approaches. Popper's approach had been entirely in terms of total *truth* content of a theory, and omitted any reference to any *similarity* between the states of affairs described by the two theories. Following the refutation of his proposal, research began aimed at understanding approximate truth in terms of similarity relations and early results were seen in the work of Risto Hilpinen (1976), and reached fruition in the work of Niiniluoto (1987) and Graham Oddie (1986), and also a collection edited by Theo Kuipers (1987). Broadly, the similarity relation is explored in terms of a metric that expresses the distance between a possible world and the actual world. But Miller, having deflated Popper's proposal, did the same for this approach by showing that if two theories are

§4 Verisimilitude, Approximate Truth, and Truth-likeness

translated into another language their relative truth-likeness can be reversed.[5] Miller's argument has never been decisively refuted and Peter Urbach successfully shows that any approach based on similarity leaves open the question – under which aspect is the similarity claimed? As he says:

> a shoe is more like a ship than sealing wax in some respects but in other ways there is greater similarity with the wax. (1983: p271)

Niiniluoto admits that:

> truthlikeness is not a purely semantic notion, but it has also a pragmatic or methodological dimension. This is due to the fact that the concept of similarity is pragmatically ambiguous. (1999: p77)

I shall return again to this question of the pragmatic relativity of approximate truth and its intrinsic difference from the concept of *truth*.

In Jerrold Aronson (1990), and Aronson, Harré & Way (1994), scientific theories are construed as 'type-hierarchies' which aim to capture the structural relationships between natural kinds. A type-hierarchy is a tree-structured graph of nodes joined by links, where nodes represent objects or concepts and links represent the relations between them. The poor progress on this proposal is perhaps indicated by Niiniluoto's understated summary of this period:

---

[5] See Miller (1976).

§4 Verisimilitude, Approximate Truth, and Truth-likeness

studies in verisimilitude have been actively continued, and interesting results and applications have been achieved, but not many dramatic novelties. (1998: p1)

In all of these works we see the deployment of much formalism and complexity but it is fair to say that none of the proposals have gained acceptance, the critical remarks of Chakravartty and Psillos (both defenders of the concept of approximate truth) being fairly typical:

> According to the 'possible worlds' approach, truth-likeness turns out to possess odd features such as dependence on the number of states of the world. True propositions end up having the same verisimilitude as false ones. (Psillos, 1999: p268)

> I cannot see how the type-hierarchical approach to scientific theories can offer a realist account of verisimilitude, and of truth (if truth is a limiting case of verisimilitude). (*ibid:* p273)

> The difficulties reviewed and suggested here for extant accounts are serious, and in some cases fatal (Chakravartty, 2007: p218)

These three major approaches to a formal analysis of approximate truth have shown little success in terms of reasonably broad acceptance. Of course, any philosophical theory always faces *some* objections, but the history of attempts at a formal analysis of approximate truth now spans almost fifty years of quite detailed work, at the end of which we see theories that have been outright refuted, others that face acknowledged difficulties, and none that gain any general acceptance, even from realists themselves, who are clearly most in need of such a formal theory. As we have seen, prominent realist advocates such as Psillos and Chakravartty openly acknowledge the failure, and Alexander Bird, another realist advocate, seems to turn away from *truth* based arguments for realism altogether when he says:

> the sorry history of attempts to characterize approximate truth does show how difficult it is to give even an informal and general account of verisimilitude, let alone a formal and general account. ... the concept of knowledge does seem to be in a better shape to do the job of forming the basis of an account of scientific progress (2007: p74)

§4.3 Psillos's Intuitive Proposal

That is the background of failure against which Psillos (1999: §11) examines all these formal approaches. He performs a service on behalf of anti-realism by exposing their deficiencies and acknowledging that none of them work. However, the title of his book promises to 'refill the realist toolbox' so he sketches a proposal for viewing approximate truth as a primitive and non-analysable concept of which we have a good intuitive awareness. He then claims that we do not need to go beyond this intuitive notion, his reason for this claim being that a formal account of *truth*, such as that of Tarski, was required because that notion fell foul of the well known *truth* paradoxes. However, Psillos claims that approximate truth is not subject to such paradoxes, and that consequently the intuitive concept is sufficient.

He is certainly right to say that 'the lack of a formal account is not necessarily a defect' (p277) and that 'we must avoid confusing clarity with formalisation' (p278). The problem is that with every attempt made by Psillos to characterise this intuitive concept, the vagueness and uncertainty remain. He first says that a theory is approximately true

§4 Verisimilitude, Approximate Truth, and Truth-likeness

if the entities of the general kind postulated to play a central causal role in the theory exist, and if the basic mechanisms and laws postulated by the theory approximate those holding in the world, under specific conditions of approximation. (p277)

Here we move from the vagueness of approximate truth to an equal vagueness concerning the phrase 'approximate those holding in the world, under specific conditions of approximation'. Despite his stated intention to offer an intuitive explication of approximate truth, his second, and more precise, phrasing begins to look more like another formal attempt:

A description D approximately fits a state S (i.e. D is approximately true of S) if there is another state S' such that S and S' are linked by *specific conditions of approximation*, and D fits S' (D is true of S'). (p277, my emphasis)

This still leaves us with that phrase 'specific conditions of approximation' as well as the suspicion of circularity since the relation 'fits' is part of both *definiens* and *definiendum*.

Psillos's account faces some severe problems, the first of which is that with approximate truth characterised as above we have a clear dependency on *truth* simpliciter, and that must entail that approximate truth is subject to those same paradoxes from which Psillos claimed it was free. However, Psillos has already said that the truth paradoxes necessitate a formal account of truth, so a formal account of approximate truth must be needed after all. Moreover, if this objection is circumvented by offering some formal account of *truth*, then it can no longer be claimed that approximate truth has been given an intuitive definition. Either way, the claim that an intuitive definition is sufficient for approximate truth cannot

§4 Verisimilitude, Approximate Truth, and Truth-likeness

be upheld so long as that intuitive definition includes the notion of *truth*.

The second issue stems from the realist's clear need to establish a connection between *truth* and approximate truth, and to this end some have suggested that *truth* is a limiting case of approximate truth (Psillos included – see his p273, where *truth* is assumed to be a limiting case of verisimilitude). But how can this be so given that the notion of approximate truth depends upon the notion of *truth*? For we cannot define approximate truth in terms of *truth* and then also claim that *truth* is a limiting form of approximate truth. If the realist replies by giving up the idea that *truth* is a limiting case of approximate truth then his position begins to look even more problematic; for, as observed above, realists are ultimately concerned with *truth*. Resorting to approximate truth was a pragmatic expedient given the untidiness of the match between scientific theory and the actual world, but surely the realist aim is *truth*. However, if *truth* is not a limiting case of approximate truth then how does the realist propose to attain *truth* via her use of approximate truth? For here she faces the problem that Psillos's intuitive account, with its 'specific conditions of approximation', ensures that approximate truth is dependent upon context. But *truth* is not supposed to be context-dependent on pain of the appearance of a kind of relativism that will be anathema to the realist. The realist is therefore left with an unbridgeable gap between the intuitively defined approximate truth and the *truth* that she ultimately seeks.

The most pressing problem, and the one I shall devote most attention to, concerns the fact that Psillos's intuitive concept cannot do the work that realists demand of it. Psillos seems to believe that anti-realist objections to the concept of approximate truth hinge on the realist's failure to give it a philosophically rigorous account, but whether the account is

§4 Verisimilitude, Approximate Truth, and Truth-likeness

formal or intuitive is not really the point – what matters is whether the concept is sufficiently precise to do the job required of it. Cheryl Misak points out that if someone

> wants to argue that our theories are getting closer to the truth, she needs to specify some traits which make one false theory more like the truth than another. (2004: p120)

John Worrall suggests that we don't even know if the notion of approximate truth is transitive, a feature surely needed by realists:

> Realists need to claim that although some presently accepted theory may subsequently be modified and replaced, it will still look "approximately true" in the light ... of the theory (if any) which supersedes the theory which supersedes it, etc. But is transitivity a property that the notion of approximate truth possesses even intuitively? (1989: pp104-5)

Surely any account of approximate truth must be of sufficient precision to answer such questions, and should also satisfy the following criteria if the concept is to be able to play any useful role in arguments for realism:

(a) It must be possible to evaluate whether or not any given theory is approximately true, or at least to state the way in which such an evaluation would proceed. This is an essential requirement for two reasons – because realism asserts that our best current theories are true or approximately true; and because the 'explanationist

§4 Verisimilitude, Approximate Truth, and Truth-likeness

defence of realism'[6] says that such approximate truth is the explanation of the success of a theory.

(b) It must be possible to evaluate whether claim RC2 (see p94) is true for a theory and its successor – i.e. does the successor of a theory contain more approximate truth? In other words there must be some means of measuring approximate truth for the convergence claim of RC2 to have much meaning.

(c) Can a theory be approximately true if its central terms do not refer, or is the answer to that again context-relative?

These are three general questions, but here are two of a more specific nature:

(d) Was Newton's theory approximately true even though its concepts of space and time were fundamentally different from those of its successor theory? How does the concept of approximate truth enable this to be answered? Could the question even have any determinate answer? Isn't the answer surely just a matter of opinion, depending on one's perspective within the debate?[7]

(e) How are the various interpretations of quantum mechanics assigned relative approximate truth values?

---

[6] A phrase used by Psillos to refer to, essentially, the No Miracles Argument – the view that realism is the only explanation for the success of science – truth explains success, and nothing else does. See §8.

[7] The context-relativity referred to earlier concerned the need for differing standards of 'approximation' in different theories. My suggestion here is that the relativism goes deeper – to the level of differing personal opinion, differing perspectives.

§4 Verisimilitude, Approximate Truth, and Truth-likeness

What is the procedure for doing this, described in a manner that is not perspective dependent?

However intuitively attractive any description of approximate truth may be, if it does not enable these questions to be answered, or at least show some method whereby they could be answered, then it is of no value to realism nor to the realism/anti-realism debate. Moreover, question (a) poses a special problem for realism for it is hard to see how (a) could be answered even in principle, regardless of what definition is used. This is because of an issue that will be discussed later in greater detail in §8.5 – the realist assumes some notion of correspondence, whether it is the correspondence account of *truth* or some notion of theory-world correspondence. As a result she has no way of ascertaining definitively whether a theory is true since there is no way to access the world other than via the theory, no *view from nowhere* standpoint from which to independently establish that *truth*. But if we cannot establish the *truth* of a theory with certainty, then how can we establish its approximate truth? The explanationist defence of realism claims that the explanation of the success of our best theories is their possession of the property of approximate truth, but we have no means, in principle, of confirming they have that property.

Let me set against Psillos's advocacy of the intuitive concept two equally intuitive examples that illustrate why it cannot offer any kind of test concerning the questions listed above: I suggested earlier that that our notion of approximate truth stems from the intuition that simple atomic propositions can provide support for such a concept. Clearly, if John actually weighs 70kgs, then it is reasonable to say that the proposition 'John weighs 71kgs' has more approximate truth than the proposition 'John weighs 65kgs'. But if, in addition, John is 180cms tall, then consider these two theories T1, T2:

§4 Verisimilitude, Approximate Truth, and Truth-likeness

Truth: John weighs 70kgs and is 180cms tall
T1: John weighs 71kgs and is 185cms tall
T2: John weighs 65kgs and is 178cms tall.

How can the relative approximate truths of these two theories be compared at all unless something additional is known about the relative importance of the height and weight parameters? Comparison of scientific theories will always involve the assessment of the relative importance of different factors – pragmatic dependence on context rears its head again. Of course, these are nothing like real scientific theories, but if this pragmatic context dependency can be seen in such a trivial case, how much more will it be present in real theories.

Against this it may be suggested that some kind of metric could provide an objective measure of the approximate truth of T1 and T2, for example the square root of the sum of the squares of the difference between each parameter and its true value, with the approximate truth measure increasing with decreasing result. This would yield in this case:

Approximate truth T1 = $\sqrt{[(70-71)^2 + (185-180)^2]}$ = $\sqrt{26}$ = 5.1

Approximate truth T2 = $\sqrt{[(70-65)^2 + (180-178)^2]}$ = $\sqrt{29}$ = 5.4

So T1 is closer to the *truth* than T2. However, this introduces many problems. To begin with, how do we choose this metric? It isn't hard to see that a different metric could be chosen which might reverse this result. Secondly, this kind of metric needs us to already know what the *truth* is, but for scientific theories, even if we could eventually know the definitive *truth*, we certainly couldn't know it in advance.

§4 Verisimilitude, Approximate Truth, and Truth-likeness

However, there is a deeper problem – as I said, the 'theories' T1 and T2 hardly resemble real scientific theories and are more like simple single propositional statements. Scientific theories will include theoretical commitments and explanatory judgements that will make the evaluation of such metrics impossible. For example, suppose that we extend T1, T2 thus:

T1a: John weighs 71kgs and is 185cms tall and this is explained by his genetic inheritance.

T2a: John weighs 65kgs and is 178cms tall and this is explained by social factors.

Again, I don't suggest that this approaches a full blown scientific theory, but T1a and T2a now begin to look more like real scientific theories which do typically include statements like 'this is explained by his genetic inheritance' that are qualitative and/or explanatory and are thus not amenable to use with a simple metric. Just as it was impossible to compare the approximate truth of two theories which assert absolute space and relative space respectively without some kind of pragmatic judgements being made, no metric could be applied to T1a and T2a without, again, pragmatic judgements being taken.

This brings me to my second example. My neighbour is in his house next door. Consider the statement:

> It is approximately true that my neighbour is in *my* house.

From the perspective of our local postman, this would be very unhelpful and he would simply say it is false. But from the perspective of someone mapping the whole of Europe, it might very well pass as approximately true. This demonstrates that approximate truth includes an ineliminably pragmatic and

§4 Verisimilitude, Approximate Truth, and Truth-likeness

context-relative aspect. In this respect it is important to note a fundamental difference between *truth* and approximate truth, for the latter has this intrinsically pragmatic component, and that is not so of *truth*.[8] This fact may undermine the possibility of any definite and determinate link between these two notions.[9]

Given the above problems, Psillos's promise to 'refill the realist toolbox' seems to amount to throwing out of the box all the somewhat rusty old tools that had been in it, and leaving instead what Laudan called a 'promissory note' – a sketch of a plan for a tool, but a sketch offering the realist little reason for hope.

§4.4 Chakravartty's Proposals

Chakravartty (2007: p214-8) also reviews the various approximate truth proposals, and, like Psillos, finds them wanting. He acknowledges the usefulness of Psillos's proposal for an intuitive notion, but raises complaints some of which are similar to the ones I make above. So he sets out to give more details of this intuitive notion, and conducts a discussion aimed at drawing out parallels between the notion of approximate truth as it applies to science and as it might apply to art, and in this he refers to Nelson Goodman's work on aesthetics. Chakravartty builds upon two notions discussed earlier in his book, namely 'abstraction' and 'idealisation' (see glossary) and he suggests that a theory's deviation from

---

[8] At least, certainly not in the non-epistemic accounts generally favoured by realists.

[9] Of course this problem is mitigated if a pragmatic account of *truth* is also embraced, though most realists would reject such a proposal, Ellis being an exception – see p49n21.

§4 Verisimilitude, Approximate Truth, and Truth-likeness

the *truth* is always in terms of these two factors. He describes these two claims:

(a) approximate truth *qua* abstraction is such that:

Approximate truth may be gauged in terms of the numbers of relevant features of their target systems they describe, so that theories incorporating greater numbers of these features are more approximately true than those incorporating fewer.

(b) approximate truth *qua* idealisation refers to:

the accuracy with which [descriptions] characterise the natures of the specific properties and relations they represent. (p228)

He finally concludes with his specific proposal for an expansion of the intuitive understanding:

Greater approximate truth can be understood in terms of improved representations of the natures of target systems in the world, and this improvement can be spelled out along two dimensions: how many of the relevant properties and relations one describes (abstraction), and how accurately one describes them (idealisation). (p230)

The analysis of the topics of idealisation and abstraction that might be appropriate here is beyond the scope of this work. Nevertheless, it is not clear that the sharp distinction between abstraction and idealisation which Chakravartty requires can be straightforwardly made, since the abstraction of parameters and their idealisation is inter-linked. As an example, consider the pendulum. Chakravartty views the removal from the analysis of air resistance as an abstraction. However, this is surely also an idealisation since that abstraction greatly

§4 Verisimilitude, Approximate Truth, and Truth-likeness

simplifies the analysis of pendulum motion. Similarly, Chakravartty takes the representation of the pendulum bob as a point mass to be an idealisation, but this can also be seen as an abstraction of some of the bob's features, for example, that it has extension, that its mass and shape may not be uniform, etc.

However, even if this doubtful distinction is allowed, I suggest that both of his definitions of these two terms still harbour considerable vagueness. Firstly, note the word 'relevant' in (a). This opens the door to all the problems that plagued the earlier similarity-based formal account – for both relevance and similarity are context-relative. What may be relevant for the realist may be irrelevant for the anti-realist – the uncertainty as to what would class as relevant features of theories would leave realists free to pick and choose to their advantage, and anti-realists similarly. Secondly, (b) leaves open the question of how the accuracy of a description is to be judged in a determinate way – see my example concerning height and weight on p107. Moreover, it seems that the realist's convergence would be guaranteed since the measure of how accurate a description is would presumably be in terms of a comparison with the presumed *true* description of our current best theories. Thus descriptive features would pass the test if they agree with current theories, but not otherwise. This is one of many examples we will see where the realist seeks to interpret the past in the light of his assumption that current theories are true or very nearly so, and that makes convergence inevitable.

Psillos may thank Chakravartty for having very interestingly expanded upon his intuitive notion, but unfortunately it still leaves unclear how the questions posed above are to be answered, specifically §4.3(d) – how does this intuitive notion enable us to determinately decide whether Newton's conception of absolute space and time counts as an approximation on the way towards Einstein's? It simply cannot do so without reference to the subjective opinions of those in the debate. That is unacceptable as it enables realists to decide the issue in their favour, and anti-realists in theirs.

§4.5 Conclusion: Approximate Truth – a Failed Research Project

For almost half a century the attempt to clarify the notion of approximate truth has continued, but without success. The work on attaining a formal definition has arguably ground to a halt and, perhaps of more significance, most advocates of scientific realism now acknowledge the failure of this project. Bird, certainly a realist, may well signal a decisive move away from any reliance upon the concept of approximate truth when he points to a problem that arises when we consider scientific progress: suppose we have a scientific theory that is approximately true, and that one new approximately true belief is added to this theory. Then this theory's approximate truth has not increased, but remained the same, for the theory is not closer to any putative final actual *truth*. But nevertheless, in this case, scientific progress may have occurred, so the case illustrates that science can progress without any increase in approximate truth. Bird's specific example is:

> There is clearly something better about believing 'the Earth's orbit is elliptical and Earth's orbit sweeps out equal areas in equal times' than just believing 'the Earth's orbit is elliptical'; but that improvement is not that the conjunction is overall closer to the truth than the single disjunct. It is that the conjunction has more worthwhile content than the single conjunct. (2007b: p75)

§4 Verisimilitude, Approximate Truth, and Truth-likeness

He claims that this illustrates the inadequacy of the concept of approximate truth regarding scientific progress:

> For we want to apply the notion of progress not simply to sequences of atomic propositions but also to sequences of complex propositions, hypotheses, and theories, to successive phases in the development of a scientific field, and even to all of science. (*ibid.*)[10]

Giere dismisses approximate truth as a 'bastard semantic relationship' (1988: p106) and Urbach concludes that

> the attempt to make sense of an objective notion of degrees of closeness to the truth for false theories is fundamentally and irretrievably misguided. (1983: p267)

Da Costa & French suggest that the absence of a formal account renders "the realist's invocation of approximate truth as so much vague hand waving" (2003: p11). Finally, after observing that 'none of the captains of the verisimilitude industry' can tell him anything about closeness to *truth* in 'less than 100 pages of complicated formulas', Musgrave says that

> the entire verisimilitude project was a bad and unnecessary idea. Popper's definition of the notion of 'closeness to the (whole) truth' did not work. The plethora of alternative definitions of 'distance from the (whole) truth' that have taken its place are problematic in all kinds of ways. (2006)

---

[10] One could also read Bird's objection as again pointing up the problem of context-relativity. For whether or not new approximately true beliefs, when added to a theory, serve to increase that theory's overall approximate truth seems to be context-relative.

§4 Verisimilitude, Approximate Truth, and Truth-likeness

Recently there have been attempts to turn to an intuitive notion of the concept. I do not think this amounts to the claim that approximate truth is a primitive and non-analysable concept, because the attempts of both Psillos and Chakravartty to flesh out the intuitive concept have distinctly formal overtones. While Psillos's discussion of the intuitive concept is little more than a gesture, Chakravartty does add a lot more detail, though his proposal, like all the others, fails a fairly simple test: Can we decide the *truth* or otherwise of the claims made by the various definitions of realism, or the claims made by the explanationist defence of realism? The answer to this question is no, and moreover, most realists now acknowledge this. Approximate truth is the Achilles heel of the realist project as there is no determinate way of assessing the claims of realism where they rely upon the concept.[11] As Laudan said:

> until the realists show us what that connection is [between approximate truth and predictive success], they should be more reticent than they are about claiming that realism can explain the success of science. (1981: p32n12)

We shall later see that a realist claim of central importance is that if a theory is approximately true, then it will be successful. This has to be regarded as a groundless claim given the complete absence of clarity or definition of what it means for a theory to be approximately true. For over thirty years scientific realism has nevertheless claimed that our best current theories are approximately true, and that the success of science is best explained by its convergence upon *truth* via a succession of theories which possess increasing approximate

---

[11] Or, as one very prominent realist said to me in private correspondence 'approximate truth is the cross that realism must bear'!

§4 Verisimilitude, Approximate Truth, and Truth-likeness

truth. Unfortunately realists continue to simply ignore this problem, as evidenced by the fact that Psillos's (1999) major defence of realism devotes the whole of chapter 11 to the problem of approximate truth without ever even acknowledging the simple fact that approximate truth does not entail success. This is a remarkable and prolonged feat of intellectual self-deception by the realist community. I suggest that it is literally a scandal that realists continue to refer to theories as being approximately true in a manner utterly central to their conception of realism but without ever having offered a theory of truth approximation that is properly related to the realist notion of truth, and is genuinely linked to theory success. Much of the problem between scientific realism and anti-realism of the COSR variety stems from the realist's reliance upon this broken-backed notion.

§5 The Pessimistic Induction Argument

§5.1 Introduction

My aim is to show that the arguments offered in support of the COSR position are unconvincing. These arguments include those which realists have levelled against the arguments for anti-realism, the most powerful and well known of which is based on the historical record of science. It is generally referred to as the Pessimistic Induction (PI).[1] In this chapter I present this argument, and in the next chapter I shall attempt to refute the many arguments which realists have brought against it. My purpose in this is not to argue for anti-realism. Rather, as with the whole of this essay, my aim is to show that the realist arguments fail. The PI refutes the realist's presumption of a link between theory success and approximate truth.[2] Central to the realist conception is the view that the success of science calls for an explanation, and that it is the truth, or approximate truth, of our best current theories that provides that explanation. This is, briefly stated, the No Miracles Argument (NMA) – given their success, if current theories were neither true, nor approximately true, that would constitute a miracle, for the only explanation of the success of science is the truth or approximate truth of its theories. Hence the argument has been called the 'explanationist defence of realism'.[3] I shall examine that argument in detail in §8, together with objections to it, but the reason for introducing it here is that it seeks to support claim RC2 (see p44) by establishing a link between theory success

---

[1] Also sometimes referred to as the *Pessimistic Meta Induction*.
[2] As previously stated, 'theory success' refers to predictive success (see p28n5).
[3] Psillos (1999: p78) characterises it as "Boyd's 'explanationist defence of realism'".

§5 The Pessimistic Induction Argument

and theory truth or approximate truth (henceforward I refer only to 'approximate truth' which can be taken as including 'or truth'). In many respects the NMA and the Pessimistic Induction (PI) argument represent the heart of the scientific realism/anti-realism debate. The NMA might seem the natural choice to be described first. However, my approach is first to describe the arguments against realism, and show that realists have failed to refute these, before discussing the NMA, an argument *for* realism, in detail in a later chapter.

Paul Horwich says that realism 'comes to grief over the history of science' (1982b: p135) and some have described the historical evidence as the greatest challenge faced by scientific realism.[4] Like so many of these arguments, it goes back a long way – certainly to Henri Poincaré, who said:

> The ephemeral nature of scientific theories takes by surprise the man of the world. Their brief period of prosperity ended, he sees them abandoned one after the other; he sees ruins piled upon ruins; he predicts that the theories in fashion today will in a short time succumb in their turn, and he concludes that they are absolutely in vain. (1905: p160)

There is indeed good reason to think that the actual historical record of scientific theory change constitutes a powerful argument against scientific realism in general, and particularly this claimed link from success to approximate truth. The PI is just such an empirically based argument, the data coming from examination of the actual history of science, for even the most cursory glance at the historical record reveals considerable ontological discontinuity in theory change, and that gives grounds for believing that in the future many of the

---

[4] For example, Worrall (1982: p216), Kitcher (1993: p136), Leplin (1997: p136).

§5 The Pessimistic Induction Argument

theoretical terms of our best current theories will be regarded as non-referring.

Thought of in that way, the PI is an inductive argument, but, as we shall see below, it can also be viewed as a deductive refutation of the realist's claimed link between success and approximate truth. However, first its inductive aspect – in some respects this was anticipated by the ancient Greek sceptics,[5] and so called 'pessimistic inductions' about theoretical knowledge are a classical form of scepticism as follows:

- *p* is currently widely believed by most experts.
- *p* is just like many other hypotheses that were widely believed by previous experts but are disbelieved by most current experts.
- Because *p* is like its predecessors, we have every reason to expect it to undergo the same fate.
- Therefore we should be at least agnostic about *p*, if not positively disbelieve it.

More recently Hilary Putnam (1978: p25) presented it in a similar inductive form:

PI1 Assume most current scientific theories are true.
PI2 So most past scientific theories are false since they were superseded by current theories.
PI3 By induction on previous theories, it follows that most current theories are false.

---

[5] See Annas & Barnes (1985).

§5 The Pessimistic Induction Argument

While this is only briefly and simply stated by Putnam, and needs more rigorous presentation, it does require a little more in the way of refutation than some realists have offered. For example, Peter Lewis (2001: pp2-3) suggests the argument is flawed because it assumes that current theories and past theories are similar, but according to Lewis they are not because current theories represent explanatory success where past theories represented explanatory failure. This seems to be rather obvious question-begging as it amounts to denying the inductive step of PI3 because of the assumption that our current theories are historically unique in having more explanatory success than their predecessors. This ignores two facts: first, that this would have been said of all previous theories; second, that we can expect that in the future our current theories will appear explanatorily unsuccessful when compared with the new theories of the future.

Because Larry Laudan's paper appeared shortly after Putnam's the name 'pessimistic induction' has stuck and many others have taken his argument as being inductive.[6] However, I think the best way to understand Laudan (1981, 1983, 1984a) on this topic is in terms of a deductive refutation of NMA and the realist's assumption of the link from approximate truth to success. For the PI argues that close examination of the history of science shows us many theories which are not only now believed to be untrue, but weren't even approximately true, and yet many of those theories were extremely successful by the same criteria of success that we apply to current theories. It would thus be a valid deduction that the truth, or approximate truth, of theories cannot be the explanation for the success of science which the NMA claims. Viewed in this way the PI argument is not so much an

---

[6] For example, Leplin (1997: p137), Herman De Regt (1994: p11), Psillos (1999: p105).

argument *for* anti-realist scepticism, but a refutation of the NMA, the principal argument for the realist position.

If the PI argument can be substantiated then it would create two difficulties for the realist. Firstly, it would show that a theory's success cannot be used to infer its approximate truth, since so many previous theories refute this. Second, it would show science to be a discipline whose history demonstrates that it is very capable of great success without any reference to truth – indeed a discipline whose historical success seems singularly unconnected with 'truth'.

§5.2 Laudan's 'Pessimistic Induction': The Argument Presented

The presentation of the argument in this chapter is relatively brief. In the next chapter I attempt to rebut all the arguments against it at much greater length and that will involve further elucidation of the argument. The two chapters stand together.

Laudan's 'Confutation of Convergent Realism' (1981) is a subtle and comprehensive statement of the argument, including considerable historical analysis. It includes a list of theories he claims to have been both successful and false (1981: p33):

§5 The Pessimistic Induction Argument

Table 2: Laudan's List[7]

- The crystalline spheres of ancient and medieval astronomy
- The humoral theory of medicine
- The effluvial theory of static electricity
- The phlogiston theory of chemistry
- The caloric theory of heat
- The vibratory theory of heat
- The electromagnetic ether
- The optical ether

This wide ranging paper, the *locus classicus* of the historical case against realism, is generally regarded as following basically the same form as the classical inductive argument given above:

- Many empirically successful theories in the history of science (such as those listed) have subsequently been rejected and, according to our current theories, their theoretical terms did not refer.
- There is no difference in kind between those discarded theories and our current theories, so there is reason to think that in the future they also will be discarded.
- Hence, we don't have reason to believe in the approximate truth of our current theories, nor the successful reference of their theoretical terms.

However, as I have said, Laudan's argument is best presented in a deductive form which would be immune to some obvious

---

[7] Note that this list is, in effect, considerably extended by those theories documented by Lyons, and listed on p178 of this essay.

§5 The Pessimistic Induction Argument

objections to such an induction. Examples of such objections would include the argument that while we may be justified in assuming that the properties of physical objects show a natural uniformity, we lack reason to assume any such uniformity in the domain of scientific theories. There is also this kind of criticism offered by Psillos:

> This kind of argument can be challenged by observing that the inductive basis is not big and representative enough to warrant the pessimistic conclusion. (1999: p105)

We can conceive of Laudan's list of older theories that were successful but false as a presentation of a phenomenon which falsifies one of the premises in the realist's 'explanationist' argument, which I take to be something like this:

EXR: The Explanationist argument for realism:[8]

EXR1: Current theories are highly successful.

EXR2: The best explanation for the success of our theories is that they are true or approximately true.

EXR3: Since EXR2 is the best explanation, we should believe it true because science shows us that such inference to best explanation leads to truth.

EXR4 Therefore we should be scientific realists.

Laudan's list stands as a factual and deductive refutation of EXR2, and that removes the motivation for EXR3, and then EXR collapses. Thus the PI directly undermines the NMA because approximate truth and successful reference cannot be

---

[8] This formulation is close to that used in §8.1 where the NMA is discussed more fully.

§5 The Pessimistic Induction Argument

the explanation of the predictive success of theories. Moreover, this conclusion is not reliant upon any statistical analysis, and is thus not prey to base rate arguments such as those discussed in Appendix 2. For as long as there are even just a small number of theories that were successful but obviously not approximately true then approximate truth cannot be the explanation of success. If there is some other explanation for the success of those false theories, then unless it can be shown that they are in some way a special case, we should assume that whatever that other explanation is, it may also be applicable to the success of our current theories. Thus EXR is defeated. The PI also represents a considerable threat to scientific realism because if current science is correct, then the ontologies of previous theories were very inaccurate descriptions of the furniture of the world. But if that is so even though they were successful predictors, then the predictive success of our current theories *does not* entail that they are correct in their descriptions of the world. As K. Brad Wray puts it:

> Given that some theories enable us to make accurate predictions despite the fact that they misrepresent the world, the fact that a particular theory enables us to make accurate predictions does not provide us with adequate grounds for claiming that it accurately represents the world. (2007: p86)

Laudan is explicit that he does not argue *for* anti-realism – '*Nothing* I have said here refutes the possibility in principle of a realistic epistemology of science' (1981: p48). However, his paper is a veritable arsenal of sceptical arguments against what he sees as the poorly made out case for what he calls

§5 The Pessimistic Induction Argument

'epistemic realism'.[9] This he characterises by five claims that he attributes variously to Putnam (1975a), Boyd (1973), Newton-Smith (1978). I reproduce those claims verbatim (1981: p21):

R1 Scientific theories (at least in the 'mature' sciences) are typically approximately true and more recent theories are closer to the truth than older theories in the same domain.

R2 The observational and theoretical terms within the theories of a mature science genuinely refer (roughly, there are substances in the world that correspond to the ontologies presumed by our best theories).[10]

R3 Successive theories in any mature science will be such that they 'preserve' the theoretical relations and the apparent referents of earlier theories (i.e., earlier theories will be 'limiting cases' of later theories).

R4 Acceptable new theories do and should explain why their predecessors were successful insofar as they were successful.

R5 Theses R1-R4 entail that ('mature') scientific theories should be successful; indeed, these theses constitute the best, if not the only, explanation for the success of science. The empirical success of science (in the sense of giving detailed explanations and accurate predictions) accordingly provides striking empirical confirmation for realism.

---

[9] He makes clear (1981: p20) that he is not concerned to attack semantic realism (that all theories have truth values), nor what he calls *intentional* realism (that theoretical terms of theories are intended by scientists to refer to existing entities). However, note my objections in §2.4 to his use of the phrase 'epistemic realism'.

[10] Laudan here takes the realist's metaphysical and semantic claims to be equivalent.

§5 The Pessimistic Induction Argument

I shall not discuss R3, R4 very much though they are referred to briefly in §5.5 below. R1 is concerned with the truth or approximate truth of theories and is identical with my RC2. R2 is concerned with the question of whether a theory's theoretical terms successfully refer and corresponds to my RC3 but with the added insistence that the reference is successful, and thus corresponds to my metaphysical claim RC1 (see §2.3.5). Clearly R1 and R2 are logically independent claims. Firstly, because many thoroughly false theories could be constructed in which R2 holds; and secondly, because it is at least logically possible that a theory could be approximately true even if some of its theoretical terms fail to refer.

However, Laudan goes on to claim, quite reasonably, that 'a theory can only be true or nearly true ... if its terms genuinely refer...' (1981: p26). Laudan doesn't argue for this, but I share his intuition – that a theory which includes non-referring theoretical terms couldn't be approximately true, but this has to remain an intuition due to the lack of clarity surrounding the term 'approximate truth'. However, the credibility of the intuition can be increased. In §5.4 I shall demonstrate that even two approximately true theories which *do* share successfully referring theoretical terms could nevertheless give wildly differing predictions, so surely any theory suffering from the additional disadvantage of having non-referring theoretical terms could not be approximately true. The difficulty of working with a term of such problematic and unclear nature as 'approximately true' is apparent here as it enables both sides of the debate to make assumptions conducive to their own positions.

Nevertheless, Laudan makes a clever dialectical move in stating the realist's position in this way. For it is not clear that realists would believe the success of scientific theories is explicable by these two factors, approximate truth and successful reference, independent of each other. It seems more likely that realists would demand the combination of both of these factors together. This interrelatedness of the two factors will come up more than once in what follows.

§5.3 Success and Reference

Nevertheless, turning first to R2, Putnam (1975a) is Laudan's main target, though he believes that the successful reference of theoretical terms is implicit in the work of other realists, and indeed, subsequent to Laudan's paper, a considerable literature emerged on this aspect of realism with realists attempting to refute Laudan's claims. I shall examine these attempts later (§6.3) and conclude that they are unsuccessful.

The realist wants to say that successful reference can explain the success of science, and that the success of science is to be explained by the fact that theories do successfully refer – i.e. that R2 is true. Laudan claims that something like the following additional claims are needed to supplement R2 if the realist is to get what he wants:

S1   The theories in the advanced or mature sciences are successful.
S2   A theory whose central terms genuinely refer will be a successful theory.
S3   If a theory is successful, we can reasonably infer that its central terms successfully refer.
S4   All the central terms in theories in the mature sciences do refer.

Consistent with the usage throughout this essay, the word 'successful' here can broadly be taken to mean that the theory makes correct predictions. S1 would be accepted on all sides, but Laudan argues that none of S2, S3, S4 are correct. One

§5 The Pessimistic Induction Argument

does wonder if he is creating an easy target in S2 as it seems barely credible that anyone would believe it, let alone a sophisticated realist. For theories can easily be constructed whose terms all successfully refer but which are either false, unsuccessful, or both. Musgrave gives a nice example:

> "Richard Nixon is tall, blonde, honest and never swears" refers to Richard Nixon all right, but it says a lot of false things about him and would be very unsuccessful in predicting Nixon-phenomena. (1988: p236)

It seems more likely that the realist would want to replace S2 by:

S2a   A theory which is approximately true *and* whose central terms successfully refer will be a successful theory.

The realist does not have to agree to Laudan's decision to separate approximate truth and reference into two separate argument strands, a tactic very much to Laudan's advantage. Nevertheless, in fairness to Laudan, the argument he offers against S2 seems equally effective against S2a. It is this: It seems that the realist is rather loose in his interpretation of what it is for a theoretical term to genuinely refer and that genuine reference is often claimed even if many of the descriptive claims the theory makes about the entities referred to are false. As Laudan says, referring to Putnam (1978: pp20-22):

§5 The Pessimistic Induction Argument

> Provided that there are entities which 'approximately fit' a theory's description of them, Putnam's charitable account of reference allows us to say that the terms of a theory genuinely refer. On this account (and these are Putnam's examples), Bohr's 'electron', Newton's 'mass', Mendel's 'gene', and Dalton's 'atom' are all referring terms, while 'phlogiston' and 'ether' are not. (Laudan, 1981: p24)

This question of what it is for a theoretical term to successfully refer will be examined in much more detail in my discussion of arguments against Laudan in §6.3. Meanwhile Laudan turns this looseness of definition to his advantage by taking it to describe the cases of several other theories which were unsuccessful but whose terms could be said to successfully refer in this weak sense that Putnam employs. Examples given by Laudan are:

- The 18th century chemical atomic theory was remarkably unsuccessful.
- The 19th century Proutian theory that atoms of heavy elements are composed of hydrogen atoms was very unsuccessful and met with many refutations.
- Early in its career, Wegener's theory that the continents are carried by subterranean objects moving across the earth was very unsuccessful, but after major modifications it became the current geological orthodoxy.

Laudan intends all of these to be examples of theories whose central theoretical terms did successfully refer. It does seem to me that the first two cases would be put aside by the realist because they will not satisfy her insistence that the historical record can only be used as evidence where 'mature theories'

§5 The Pessimistic Induction Argument

are concerned (see §6.4). Nevertheless, the Wegener continental drift example seems immune to this objection since its predecessor was surely a 'mature' theory. Moreover, this example seems to count against S2a just as much as S2 because the realist would presumably say that the original formulation of Wegener's theory was approximately true. Since its central theoretical term – the tectonic plate – must be taken as successfully referring, then S2a cannot be true.

Laudan concludes:

> The realist's claim that we should expect referring theories to be empirically successful is simply false. (1981: p24)

Laudan next turns to the claim S3, and here the evidence of history does seem appropriate, and seems to show many highly successful theories whose central terms, in retrospect, we no longer believe successfully referred. Theories related to the ether and to phlogiston are cited as examples.

Laudan argues that S3 is too weak. For remember that S3 enables only a 'reasonable inference' from theory success to theory reference. Hence

> If, as (S3) allows, many ... referring theories can be unsuccessful, how can the fact that a successful theory's terms refer be taken to explain why it is successful? ... it [S3] arguably gives the realist no explanatory access to scientific success. (1981: pp25-6)

Laudan complains that S3 needs the strength of an entailment from successful reference to theory success, but is this another example of Laudan's strategy of using the separation of approximate truth and reference to create straw men that are

§5 The Pessimistic Induction Argument

conveniently easy to knock down? For just as S2 was augmented to S2a we could similarly augment S3 to S3a:

S3a If a theory is approximately true and successful, we can reasonably infer that its central terms successfully refer.

Now Laudan's complaint would seem less obvious. Nevertheless, we can legitimately question whether S3 is true – does history show that successful theories have been referring theories? Does the success of a theory warrant the conclusion that its central terms refer? This is an empirical question – have past successful theories been ones whose central terms referred? Laudan answers it with many examples of highly successful ether theories that we now think did not refer. He refers to the theories of electric fluid, caloric ether, optical ether (Fresnel's 'bright spot' prediction), etc and says that in this period no theories

> were as successful as aether theories; compared to them, 19th century atomism (for instance), a genuinely referring theory (on realist accounts), was a dismal failure. ... non-referring 19th-century ether theories were more successful than contemporary, referring atomic theories. (1981: p27)

He also recalls the famous remark of James Clerk Maxwell, to the effect that 'the aether was better confirmed than any other theoretical entity in natural philosophy' (*ibid.*) So here are successful theories which, even if regarded as approximately true, included non-referring central terms.

Clearly, with both S2 and S3 refuted R2 has no support at all.

§5.4 Success and Approximate Truth

Turning now to the claimed link between success and approximate truth, Laudan (1981: p30) suggests that what the realist really wants is this pair of claims:

T1   If a theory is approximately true, then it will be successful.

T2   If a theory is successful, then it is probably approximately true.[11]

Now the following is true:

T1a   If a theory is true, then it will be successful.[12]

But of course, realists do not think we can assume that any scientific theory is true simpliciter since even our best theories may require some potential modification. Moreover, if the realist could explain only the success of theories which are true simpliciter, that wouldn't give him very much concerning the success of all the other approximately true theories.

---

[11] Actually Laudan here uses the phrase 'explanatorily successful' but it is not clear why he refers to explanations at this point. Elsewhere he simply says that 'a theory is successful if it makes substantially correct predictions, if it leads to efficacious interventions in the natural order, if it passes a battery of standard tests.' (p32). Accordingly I drop this reference to explanatory success as it seems redundant and confusing.

[12] However, Lyons (2002: p75) questions even this on the grounds that a theory being true entails nothing about the theory's success until the auxiliary assumptions are taken into account. So it seems that an entailment from theory truth to success demands the assumption of the truth of all auxiliary assumptions as well. Only in that case will a theory's truth entail its success, and hence explain it. I press this point no further other than to note this additional complexity which both realists and Laudan ignore.

§5 The Pessimistic Induction Argument

Laudan's scepticism insists that the truth of T1a does not of itself entail the truth of T1 – that needs independent argument, which the realist does not supply. However, he claims, most realists appear to use the obvious truth of T1a as license to take for granted the truth of T1. Laudan here claims that none of the writers he is aware of have given any definition of what is meant by 'approximate truth', and with that I obviously agree, but I set aside that issue for the moment.

At this point in Laudan's argument he grants, for the sake of argument, that T1 is true – if a theory is approximately true then it will be (at least somewhat) successful. But I want to go further than Laudan here and make a few observations that suggest there are strong reasons for doubting this. Musgrave (2006) points out that 'near-truths yield falsehoods as well as truths'. Presumably a theory saying there are seven planets has more approximate truth than one saying there are twenty seven, but neither is likely to exhibit much empirical success. As we have seen in our analysis of the concept, approximate truth is not a straightforwardly semantic property like truth, but also has a pragmatic aspect, so to say that a theory is approximately true has considerable vagueness and is highly context dependent. Consider the obvious case of the science of weather, or more generally, chaotic systems. Clearly here T1 is certainly not true. Someone might object that typical successful theories of physics are not like these cases, though that would be to take an 'idealised laboratory' view of those theories, and Cartwright's thousand dollar bill on a windy day in St. Stephen's Square comes to mind as counter to this view (1999: p27). However many approximate truths are known concerning that dollar bill, predictive success is not likely. Moreover, consider Newton's inverse square law, perhaps a paradigm example of the kind of physical law where intuition would declare T1 to be true. It can be shown mathematically that the inverse square law of attraction not only yields elliptic planetary orbits, but that these orbits are stable – minor

§5 The Pessimistic Induction Argument

deviations from the elliptic orbit are rapidly corrected and the planet pulls back to its correct orbit. It can also be shown that for any value of the inverse power other than two, this will not be so. Thus if the law of attraction was this:

$$F = G \times \frac{M \times m}{D^{2+\delta}}$$

then, even for very small values of δ, the orbit would be unstable – the smallest perturbation and the planet would fly off into space. So if Newton's theory had been slightly in error and proposed a law as above but with $\delta = 0.001$, his theory would have been wildly unsuccessful in its predictions, but, one assumes, would still have been approximately true.

An even more powerful example[13] concerns the theoretical charge on the electron. A very slight change in a theory's theoretical claims can lead to radically different empirical predictions. Suppose we have two theories:

TP1: The entire corpus of contemporary physics.

TP2: Identical to TP1, except that the charge of the electron is higher by one billion-billionth of its value in TP1.

Presumably TP1 and TP2 are regarded as approximately the same theory, and by any conceivable definition of 'approximate truth', if TP1 is regarded as either true or approximately true, then TP2 must surely be approximately true. Yet according to TP2 the world as we know it would not exist, illustrating how minor theoretical differences can lead to radically different empirical predictions, even outside of the context of 'chaotic' systems. TP1 is a successful theory, but

---

[13] Taken from Lyons (2002: p75).

§5 The Pessimistic Induction Argument

TP2 is approximately true but highly unsuccessful, so how can there be an entailment from approximate truth to success? This example involves two theories having all the same referring terms. How much more empirical deviation may be expected with two approximately true theories with at least one differing theoretical term? This argument can be refuted only if a definition of approximate truth can be given that enables TP1 to be approximately true while TP2 is not. No such definition has been offered so I would go further than Laudan and argue that there really is no good reason to assume the truth of T1 – approximate truth does not entail success.

This still leaves open the question of T2 – that explanatory success can be taken as warranting approximate truth. Laudan claims T2 is false. Examination of the history of science shows that there are many theories which have been highly successful for long periods of time but which have not been even approximately true in terms of the claims they made about the world. For example, Newtonian optics predicted a wide variety of phenomena but was based upon an ontology of light which we now believe to be completely wrong. There seems to be nothing which might even approximately correspond to that theory's light corpuscles, so it is clear that Newton's theory could not have been approximately true. Of course here I am again forced to make assumptions as to what 'approximately true' might mean. In particular, I am assuming that it includes a demand that the theory's central terms do all successfully refer.[14] On that assumption, it is surely reasonable to say that a theory whose central ontology was wrong cannot be considered approximately true. How then

---

[14] Laudan would agree, saying: 'the *realist would never want to say that a theory was approximately true if its central theoretical terms failed to refer*. If there were nothing like genes, then a genetic theory, no matter how well confirmed it was, would not be approximately true' (1981: p121, original emphasis).

can the realist explain the success of Newtonian optics? The realist cannot even show that an approximately true theory must be successful, so how could she explain the success of a theory like Newton's which is not even close to the truth? Laudan argues that this goes for most theories in the history of science. He famously lists many very successful theories that were based on what we now believe to be incorrect models and structures (see this document: p121) and the realist cannot hope to explain the empirical success of these theories in terms of the approximate truth of their claims.

Laudan's battery of sceptical arguments severely undermine both T1 and T2 and thus sever the realist's presumed link between the success of a theory and its approximate truth. As he says, the realist seems to have no answer to

> the prima facie plausible claim that there is no necessary connection between increasing the accuracy of our deep-structural characterizations of nature and improvements at the level of phenomenological explanations, predictions and manipulations. (1981: p35)

Theories which are approximately true need not be successful, and many theories which had great success were not approximately true. Empirical success doesn't seem to warrant approximate truth. This also vindicates the claim made in §3.7.4 that realists have failed to establish a link between scientific method and theoretical truth.

§5.5 Convergence

A major part of Laudan's paper is concerned with what he terms 'convergent realism'. For Laudan this goes well beyond the view that successive theories converge upon the truth, which is more or less my claim RC2. For, in his R3 and R4 he greatly expands upon the relatively simple idea that successive theories come closer to the truth. He thinks that a requirement is placed upon a theory, but should not be, that it must somehow explain the success of its predecessor theory, and also show the way in which that predecessor is a limiting case of itself. I do not propose to pursue Laudan's stronger notion of convergence, partly because it is not claimed by the various formulations of realism I have examined. However, it is useful to note the two conclusions he takes himself to have demonstrated:

- He follows Kuhn and Feyerabend in questioning why a test for the acceptability of a theory is that it must explain why its predecessors succeeded or failed.

- Once a theory has been falsified, it is unreasonable to expect that a successor should retain either all of its content or its confirmed consequences or its theoretical mechanisms.

Clearly Laudan here points up some assumptions as to the nature of science and its theories which flow directly from the realist conception of science.

§5.6 Conclusions

Laudan's presentation of the PI covers considerably more than is mentioned here. However, even from this brief overview, together with the additional arguments I have presented, I take this discussion to have shown:

§5 The Pessimistic Induction Argument

LC1 There is no inference from the success of a theory to the fact that its central terms successfully refer (S3 refuted).

LC2 There is no inference from the success of a theory to the fact that it is approximately true (T2 refuted).

LC3 The realist's claim that theoretical terms within the theories of a mature science genuinely refer (R2) lacks any argumentative support.

LC4 There is no evidence supporting an inference from the approximate truth of a theory to its success (T1 shown to lack credibility).

LC2 and LC4 together completely sever any link between predictive success and approximate truth, even if that term had clear and determinate meaning. We shall later see that this is a major reason to reject the NMA, which is the realist's prime argument in favour of RC1. In addition, recalling claim RC2:

RC2 Scientific theories are typically approximately true and more recent theories are closer to the truth than older theories in the same domain.

In view of these problems it is hard to see what this COSR claim amounts to. Clearly RC2 represents the realist view of scientific progress as convergence upon truth. I suggest that scientific progress is far better seen as the increased ability to make accurate predictions which facilitate useful interventions in the world. However, the main conclusion of this chapter is that the realist's claimed connection between approximate truth and explanatory power lacks credibility.

§6 Defending the Pessimistic Induction

§6.1 Introduction

Having briefly described the PI argument, I shall now attempt to rebut the many attempts that have been made to refute it. The literature here is vast, in recognition of the central importance of the PI argument against realism, and consequently this is a lengthy chapter, though the arguments fall into three separate groups as follows:[1]

(i) The first group all aim to show that the PI argument is just plain wrong in various different ways and that it fails to break the link between success and approximate truth and/or reference. This group is discussed in **Error! Reference source not found.** and §6.3.

(ii) This second group accepts the force of the PI argument, but replies by complaining that the PI conception of 'scientific theory' is cast too wide and that its argument fails once the claimed link between success and approximate truth is restricted to a subset of previous theories which can be regarded as, in some sense, 'properly' scientific. Examples are those belonging to the 'mature sciences' or those whose success involves 'novel predictions'. Once these restrictions are added, it is claimed, all (or most) of the entries in Laudan's list (p121) can be removed. This group is discussed in §6.4 and §6.5.

(iii) This third group again accepts the force of the PI argument, but replies by claiming that it was always a

---

[1] I have relegated discussion of the 'Base Rate Fallacy' argument to Appendix 2 because I do not think it is conclusive in the NMA/PI debate. Indeed, particularly in the case of the PI, I think it misses the point.

mistake for realism to have claimed a link between success and the approximate truth of the *whole* theory, for the link is actually between success and the approximate truth of some *part* of the theory – that part which was responsible for the success. Moreover, that part of the theory will also be found to show continuity across theory change, thus supporting the claimed convergence of theories. This group is discussed in §6.6.

Groups (ii) and (iii) can be seen as attenuating the realist claim from the original:

> Success of a theory entails its truth.

to an eventual:

> Within mature theories, novel success entails approximate truth of selected parts of the theory.

In the ultimate attenuation within group (iii) – Structural Realism – the truth of all theoretical commitments is lost.

§6.2 The Link Between Approximate Truth and Success

I shall first examine objections to the attack made by the PI on the realist's presumption of a link between approximate truth and success. Recall that both of these realist claims have been undermined (see §5.4):

T1 If a theory is approximately true, then it will be successful.
T2 If a theory is successful, then it is probably approximately true.

§6.2.1 Laudan's Assumption: No Reference Implies No Approximate Truth

In the case of T2, the argument was that the historical record shows large numbers of theories that have been successful but are not now believed to be even approximately true because it was insisted that any theory whose central theoretical terms do not refer can't be claimed to be approximately true. As I observed, Laudan (1981: p33) does explicitly assume that realists 'would never want to say that a theory was approximately true if its central theoretical terms failed to refer', and on p125 I presented additional arguments for the plausibility of this view. However, Hardin and Rosenberg (1982) claim that classical Mendelian genetics shows Laudan to be mistaken. They say it would be judged to have been approximately true, but its central terms failed to refer because, they claim, contemporary genetics does not recognise anything like a 'Mendel gene'. Once again the debate is afflicted by the absence of clarity of the term 'approximate truth', enabling it to be interpreted to suit different viewpoints.

However, even if this problem is set aside, this objection to the PI doesn't seem very effective, as surely the Mendel case is not typical of most theories on Laudan's list. Even if we grant the somewhat implausible proposition that contemporary genetics includes nothing like a 'Mendel gene', and that, consequently, the case could show how one specific theory could be both non-referring and approximately true, this is surely a one-off case. For this would not generally be the case given what most of our contemporary theories would say concerning their predecessors on Laudan's list. By the standards of our current theories, which the realist claims are true, or approximately so, I can insist that many successful theories on Laudan's list were neither true, nor approximately true. Examples would be: phlogiston theory, caloric theory of heat, the theory of light as a wave motion propagated through a material ether. Thus the relationship which Hardin and Rosenberg claim exists between Mendelian and contemporary genetics would not be at all typical of the relationship between theories on Laudan's list and current theories in general.

Moreover, given the examples I gave in §5.4 of correctly referring, approximately true theories giving wildly unsuccessful predictions, if such theories could also be not even correctly referring but still deemed approximately true, then the credibility of the term 'approximate truth' becomes deeply strained! I think Hardin and Rosenberg make little headway here. Given the entirely unclear notion of 'approximate truth', Laudan seems entitled to claim that whatever that notion may signify, in general, non-referring theories cannot have it.

§6.2.2 McAllister: Laudan's Examples were not Actually Successful

Another attempt to block Laudan's attempt to sever the link between approximate truth and success comes from James McAllister (1993). His strategy is to deny that, by our standards, previous discarded theories were successful, and thus reduce the length of Laudan's list. Indeed, if completely successful he would remove the list entirely. He argues that theory choice and the study of the properties of theories is now vastly more sophisticated than it was. Laudan's definition of success included 'substantially correct observations' and 'leads to efficacious interventions in the natural order' (1984b: p109). McAllister is sceptical about the latter on the grounds that this would not normally be a realist's criterion of success, and that it introduces large questions as to how we judge whether historical theories did possess this virtue. He undermines the former by suggesting that at each point where there has been radical theory

§6 Defending the Pessimistic Induction

transition it has been accompanied by a considerable upgrading of the accuracy requirements of empirical observations. In short, whilst former scientists have considered their theories observationally successful by their standards, they would not be by ours. McAllister concludes:

> the judgments made in the remote past about the [empirical success] measures of theories are in general not as reliable as those which take account of the later discoveries about the properties of theories. ... Therefore, the theories deemed successful in the history of science were deemed to be so on the basis only of a set of criteria constructed in the light of imperfect knowledge about the properties of the properties of theories. (1993: pp211–2)

He refers here to the 'remote past' and he also refers to a time when theory choice was subject to the requirements of 'consistency with the bible' (p211). These suggest that he is referring only to theories of the very distant past and he offers no reason to doubt the observational accuracy of theories of, for example, the nineteenth-century. In addition, there is a *tu quoque* argument here – the theory which realists most like to quote as an example of novel predictive success is Einstein's General Relativity, yet the controversy over the 'observational accuracy' of Eddington's 'confirmation' is well known.[2] Presumably General Relativity is a definite part of modern science, and thus McAllister's claim of stricter 'accuracy requirements of empirical observations' must either be dropped or he may have to question the acceptability of some of our most highly regarded current theories.

---

[2] See, for example, Collins & Pinch, 1998: §2.

§6 Defending the Pessimistic Induction

This argument has gained little acceptance, though, as we shall see, Kukla (1998) is an exception. I am sure that some theories on Laudan's list would be excluded by McAllister's more stringent criteria of observational success. However, the higher standards imposed by contemporary science on the criteria for empirical success can surely be satisfied by *some* past theories; for example, the classical electromagnetic theory, or the steady state universe theory of Fred Hoyle and Hermann Bondi. Thus, McAllister could not exclude the probability that some theories on Laudan's list are successful even by today's standards and thus some of Laudan's theories could certainly survive McAllister's argument.

§6.2.3 Perhaps Our Current Theories are True

In my discussion of the PI we saw how Laudan took for granted that most realists 'are (rightly) reluctant to believe that we can reasonably presume of any given scientific theory that it is true.' (1981: p30) He takes it for granted that we have good reason to think that even our most empirically reliable current theories about the world will need some modifications, so none can be true in every respect. Thus the realist is forced to fall back on the ill-defined approximate truth. But Kukla takes issue with this, suggesting that here the anti-realist is begging the question against the realist because the only reason we have for thinking current theories are false is

> Putnam's (1978) 'Disastrous Induction': all theories of the past are known to be false, so it is overwhelmingly likely that our present theories will turn out to be false as well. (1998: p16)[3]

---

[3] Kukla is a little disingenuous here since he knows that Putnam's very briefly stated induction is not what fuels scepticism, but Laudan (1981) which I have stressed need not be seen as an inductive argument at all.

§6 Defending the Pessimistic Induction

He concedes that the PI does offer *some* evidence that our current theories are false, but reasonably complains that any simple numerical induction like that of the PI must be treated with caution, and thus that the falsehood of previous theories, *by itself*, would be a weak basis for concluding the falsity of current ones, adding that

> our interim theoretical accounts of the world might nevertheless all be false until we got very close to the end. (*ibid.*)

The thrust of Kukla's argument here is that if he can show that at least the majority of current theories are true, then the realist does not have to fall back to claiming they are approximately true, and thus Laudan's attack will not succeed as it concerns the link between success and approximate truth – no one doubts that truth simpliciter could explain success.

Against Kukla, it seems intuitive that our current theories are not all true simpliciter, for that would signal the end of science. Physics, for example, would have reached that 'true and complete' stage that many philosophers have incorporated into their theories while not entertaining much hope that it had already arrived.[4] Is Kukla's proposal even remotely credible? It would imply that most branches of science have reached a terminus such that that further work is futile because they have entered a new phase where all that remains is the dotting of i's and crossing of t's. Even if a stage was reached where some theory was in fact the literal truth, we could have no way of knowing that was the case. Thus work would continue within that domain of science and the theory could eventually

---

[4] David Armstrong, for example, says that by 'physical properties' he is referring to 'whatever set of properties the physicist in the end will appeal to' (1991: p186).

§6 Defending the Pessimistic Induction

be changed slightly in such a way that it would then deviate from the truth. This thought, incidentally, works against the entire notion of theories being convergent on the truth in any sense.[5]

One wonders how it would help realism if current quantum mechanics, with its several different interpretations, turns out to be the terminus of theoretical exploration and is literally true. Indeed, what would it even mean for it to be 'true' since we could not know to which of those many different interpretations the word 'true' is to be applied.

Moreover, Kukla's argument would need the majority of current theories to be true, for even if some theories were true, if a large number of other theories were false but successful then we would be again entitled to believe that theory truth cannot be the explanation of theory success. Kukla acknowledges that he needs McAllister's thesis (§6.2.2) – that previous false theories were in fact not successful in order to overcome the force of the PI, and I have already argued that McAllister's thesis is incorrect.

Kukla requires that all the theories of the past be regarded as unsuccessful (in line with McAllister's argument), *and* that most current theories are true, and if both these conditions hold then the PI argument will fail. However, the total structure of Kukla's version of the explanationist defence of realism now seems strange. His full argument can be found on p246 of this essay, where it is used as the basis of my presentation of the NMA. It includes the premise that 'The

---

[5] Of course it is true that attaining a theory that was literally true wouldn't mark the end of possible useful development in that field, for it would remain possible that further new truths could be added to the theory, or its scope could be expanded. However, while these positive developments are possible, it is also possible that further development could lead the theory away from that 'truth' which had previously been attained.

only (or best) explanation for this success is the truth (or approximate truth) of scientific theories,' and reaches the conclusion 'Therefore we should be scientific realists', meaning, of course, that scientific theories are true. Now, in order to protect this argument against the attack offered by the PI, he finds it necessary to assume that current theories are true! By assuming current theories are true, his explanationist argument that they are true can be upheld!

Kukla makes an interesting proposal, but I think it fails, though he is not alone in suggesting that current theories are true. Bird, for example, questions why

> we think that our theories are not strictly correct? ... There are many venerable scientific propositions that have never been falsified and which we have no reason to suppose ever will be, and which do not state approximations. (Bird, 2007b: p73)

He goes on to give a list of such propositions:

a) blood circulates pumped by the heart
b) Chemical substances are constituted by atoms
c) Water is a compound of hydrogen and oxygen
d) Light is electromagnetic radiation
e) Electrons are negatively charged
f) The speed of light is constant for inertial observers
g) Smoking causes cancer
h) The tides are caused by the gravitational influence of the moon
i) The continents have moved over time
j) Mankind has evolved from ape-like ancestors
k) DNA has a double-helical structure.

§6 Defending the Pessimistic Induction

It seems strange to offer a list of allegedly true scientific *propositions* as evidence for the truth of scientific theories. However, setting that aside, isn't Bird taking a big risk, suggesting that, for example, (d), (f), (j) are *literally true*? Or that (f) and (k) could never turn out to be only approximations? (a) is a simple statement of observation, and so is (i) really, as it has been verified through observation. So we are left with (b), (c), (e), (h). However, subject to their different understandings of some of the terms, most anti-realists could happily accept these – (b) and (c) are simply statements of the atomic theory – an instrumentalist could accept them, though his understanding of the word *is* would be different from that of the realist. Constructive empiricists and followers of Arthur Fine's 'Natural Ontological Attitude' (NOA) [6] would similarly accept (e) – whatever is meant by 'electron' and 'charge' – (e) is true. It might be replied that what anti-realists make of (b), (c), (e), (h) is irrelevant – what matters is that they are literally true and thus exemplify true scientific propositions. But why should the anti-realist accept that these propositions are true in the realist sense? She can reasonably enquire how it is known that these propositions are literally true as opposed to being fallible propositions, or propositions that are true relative to the theory in which they arise. An anti-realist may ask how it is known that 'electrons are negatively charged' in any sense other than relative to the theory in which the words 'electron' and 'charge' gain their meaning. Moreover, if observation bears out the regularity of tides and lunar phases, then (h) must be true. The only one in the list that seems entirely non-controversial is (g) since it is clearly true that smoking is a causal factor, even if others are discovered, though it is unclear what bearing that might have

---

[6] Proposed by Arthur Fine (1984, 1986b) as a *via media* between scientific realism and anti-realism, accusing both of those positions of adding a redundant philosophical layer on top of the *natural attitudes* of scientists themselves.

here since it is merely a truth concerning a statistical connection between two completely non-theoretical phenomena which doesn't need to stand within any theory in order to remain true.

None of these are complete theories, and, in fairness, that is what Laudan's PI applies to. Doubtless a PI could be constructed concerning various statements of science, but we really are here concerned with theories. Moreover, asserting that all of these statements are literally and unmodifiably true, strikes me as a case of realist table-thumping rhetoric.

§6.2.4 Summary: Approximate Truth and Success

The PI arguments of §5.4 undermine the link between approximate truth and success. None of the arguments considered here succeed in restoring that link in either direction and we are left with no reason to believe these claims to be true:

T1 If a theory is approximately true, then it will be successful.

T2 If a theory is successful, then it is probably approximately true.

Approximate truth cannot be the grounding of a theory's predictive success since large numbers of false theories have been successful.

§6.3 The Link Between Reference and Success

§6.3.1 The Causal Account of Reference

Perhaps the most important realist replies to Laudan's historical challenge try to show that he was wrong to say that the central theoretical terms of previous successful theories

§6 Defending the Pessimistic Induction

did not successfully refer, that he exaggerated the extent to which the central terms of past theories must be regarded as non-referring. For Laudan it was not enough that a central theoretical term should refer to *some* item. He also demanded that the descriptive claims made by theories about that item must have some accuracy before we could legitimately claim that those theoretical terms successfully refer. I already noted in §5.3 that Laudan did trade upon a looseness in the realist's definition of the word 'refer', so perhaps it should be no surprise that now, in reply to him, those realists seek to clarify their meaning in a way that differs from Laudan's understanding of the term! Hence alternative accounts of the reference of theoretical terms have been prominent in attempts to reply to the PI. In §5.1 we saw that Putnam's version of the PI was a simple induction aimed at showing that our current theories are false – in other words, that the items referred to in those theories do not exist. The version I presented, following Laudan, aimed for an epistemic conclusion – that we lack reason to believe our theories to be true. We should know by now that no such conclusions can be overturned by semantic considerations. Theories of reference are precisely that – semantic considerations, and thus it should not surprise us that no amount of argument based upon theories of reference can undermine the conclusions of the PI, and this is what I will argue in what follows.[7]

Some critics of the PI point out that a causal account of reference would not demand Laudan's accuracy of description. On a causal account of reference a term *refers* to some item in the world not because the item satisfies some description associated with the term, but because of a causal

---

[7] In this respect see Bishop (2003) and Bishop & Stich (1997) who deplore the phenomenon they refer to as 'The Flight to Reference' in which substantive philosophical discussion is replaced by discussions of theories of reference.

§6 Defending the Pessimistic Induction

relationship between the item and the speaker who first introduced the term into the language.

In the case of so-called 'natural kind' terms, for instance, causal theories of reference invite us to imagine a speaker in the distant past coming into direct causal contact with an actual sample of a yellow metal substance, pointing it out to other members of her linguistic community, and announcing that she will name this substance 'gold'. On a causal theory of reference, a 'natural kind' term like gold undergoes some kind of initial 'naming baptism' which fixes the reference of the term by associating it with the underlying constitution of gold, even though that is unknown at that time. So when that linguistic community use the term 'gold' they manage to refer to the stuff with atomic number 79, even if all other aspects of the descriptions they associate with the term 'gold' turn out to be wrong. This would explain how historic language users could use terms like 'gold' to make both true and false claims about the actual stuff gold, rather than just making true claims about entities that existed only in their heads. Likewise, the realist argues, the term 'chromosomes' referred to chromosomes even when used by early theorists who had false beliefs about the nature and properties of chromosomes.

Of course, a theoretical term in a scientific theory could not have been introduced into a language by pointing out samples of their intended referents, but causal theories of reference claim that such a theoretical term refers to whatever items in the world cause the phenomena that led past theorists to introduce it into their theories. Thus, when Ampère used the term 'electricity' it referred to the *actual* electricity causing the phenomena (sparks, lightning, etc) that had originally led them to propose a theoretical item responsible for these phenomena, despite their false descriptive beliefs *about* electricity (e.g. that it was a fluid substance).

§6 Defending the Pessimistic Induction

Before examining some specific objections to the PI based upon theories of reference, I want first to raise some general doubts concerning the use of the causal theory of reference in this context. To begin with, the previous paragraph describing how theoretical terms are accommodated within causal reference theories makes it clear that a causal theory of reference already presupposes the reality of *some* item that, as it were, grounds the use of the theoretical term. Of course such theories have to allow for the fact that a particular theoretical term may fail to refer, but the core assumption is that it *may* refer – in principle, there *could* be something for it to refer to. However, that is one of the assumptions which some anti-realists reject! Thus all arguments against PI that utilise a causal theory to account for the reference of theoretical terms are question-begging against anti-realism before the rest of their arguments even get going. After all, the debate between realist and anti-realist concerns precisely the question of whether there is good reason to believe that the theoretical terms of successful theories do refer to actually existent entities, properties, or processes. However, the causal theory of reference, as it applies to the theoretical terms of successful theories, explicitly assumes that they do refer. Putnam (1975b: p274) said that the term 'quark' could have been introduced by the reference-fixing causal description 'particle responsible for such-and-such effects'. But an instrumentalist will probably deny any reference to an existent particle capable of having causal effect, and might maintain that a theoretical term is little more than a variable in a set of equations; and a constructive empiricist would at least withhold assent and maintain an agnostic position. Thus the causal theory of reference is seen to be part of a wider metaphysical view of the world which presupposes some form of scientific realism. Now it is true that the PI argument is not specifically anti-realist, being only explicitly sceptical of the explanationist defence of realism. Nevertheless, one might complain that the causal reference theorist begs the question

§6 Defending the Pessimistic Induction

that theoretical terms do refer, and that is, at least in part, precisely what is at issue.

In addition, it is not clear that scientific realism itself should be content to lose the descriptive meaning associated with theoretical terms. Consider this recent definition of scientific realism by Devitt:

> Most of the essential unobservables of well-established current scientific theories exist mind-independently and *mostly have the properties attributed to them by science*. (2004: p102; my emphasis)

This clearly represents a stronger ontological commitment than the purely referential approach in which theoretical terms just refer to *something* devoid of description. It seems plausible to insist that an essential part of any scientific theory is not simply reference to theoretical items, but also some description of what these items are. For example, the claim that light is transmitted as waves in the ether is empty without some descriptive account of what the ether is. The discussion below of replies to PI by Kitcher and Psillos will show a progressively increasing awareness of the need for descriptions that augment causal reference.

Another problem concerns the question as to whether any causal theory of reference can be correctly applied to the reference of scientific theoretical terms. This is because of a problem I call the 'ubiquity of reference' – it will always be possible to say that any theoretical term refers. Even if the scientific theory has been abandoned, every theoretical term must successfully refer, no matter how mistaken the descriptions associated with it, given that *some* thing or other was present in the grounding of the term. As a result, pure causal theories cannot explain why there would ever be any referential failure.[8]

Having stated these strong general reservations concerning the legitimacy of the causal account of reference in this context, let us examine replies to Laudan that do utilise that theory.

§6.3.2 Hardin and Rosenberg: Pure Causal Reference[9]

Hardin and Rosenberg (1982) was one of the earliest replies to Laudan, and his reply to them (1984a) effectively refuted their objections. They argued (1982: p613) that because 'One permissible strategy of realists is to let reference follow causal role', realists can regard the central terms of theories which were even radically mistaken to have been referential. For example, we regard the electromagnetic field as playing the causal role attributed to the ether by earlier theories, and thus 'It seems not unreasonable, then, for realists to say that "ether" referred to the electromagnetic field all along' (*ibid:* p614).

However, this response to the PI leads realism into trouble on two counts. Firstly, there is the ubiquity of reference problem – as Laudan himself points out (1984a: p160) it is possible for most thoroughly wrong theories to claim some kind of successful reference. Laudan gives the example of 'Cartesian vortices' which could plausibly be claimed to have been referring to gravitational attraction (*ibid.*). As stated above, a purely causal theory of reference ensures that nearly all theoretical terms could be said to refer, no matter how wrong or unsuccessful the theory may be.

---

[8] Psillos (1997: p269) points out that the converse problem arises for descriptivist theories of reference, where it is difficult to show that any term could ever successfully refer.

[9] Arguments against Hardin and Rosenberg along similar lines to those presented here can be found in Worrall (1989: pp116-7), Psillos (1999: ch.12), Stanford (2006: pp148-9).

§6 Defending the Pessimistic Induction

Secondly, and more importantly, while Hardin and Rosenberg secure a history of successful reference for terms in previous theories, they do so by driving a wedge between the reference of the terms and the accuracy of the beliefs that were part of the theories. However, if the central terms of past theories are referential *despite* the fact that the theories in which they figured were all radically misguided, then what reason have we to believe anything told to us by current successful theories? Yes, there are 'electrons', but the PI tells us that we need not believe anything science tells us about them as there is a long record of successful theories with referring terms where the beliefs associated with those terms were completely wrong.

If the argument of Hardin and Rosenberg is sound it would be rather bad for scientific realism, because we could then use the PI to show that we would have no reason to trust our best scientific theories. However, trusting the accounts of current theories is just what the realist says we should do. Perhaps Laudan deserves the last word here:

> having conceded that genuineness of reference is not a *sine qua non* for empirical success, Hardin and Rosenberg no longer have a license for treating *our* theories as any more likely to be referentially sound than those of Mendel. (1984a: p161)

I believe that the Hardin & Rosenberg paper demonstrates that a purely causal account of the reference of theoretical terms will not do for scientific theories. Some aspect of the meaning of those terms must surely be maintained. Such meaning would include the descriptive claims made by scientists concerning those terms, and also the meaning the terms take from their overall place within the theory.

§6.3.3 Kitcher: Causal Reference *or* Descriptions

After Hardin and Rosenberg's early attempt to undermine Laudan's paper it was over a decade before the next significant reference-based attempt, found in the arguments of Kitcher (1993: ch.5). This may reflect the decisive refutation which Hardin and Rosenberg received. Nevertheless, whilst Kitcher's arguments involve a sophisticated new proposal for how theoretical terms refer, I will argue that the main lesson doesn't seem to have been learned from the mistakes of that earlier paper. For Kitcher's arguments would lead to the same loss of trust in scientific theories as we found in Hardin and Rosenberg. Kitcher's proposals have been criticised even by fellow realists such as Psillos (1997) and Christina McLeish (2005) and my analysis echoes some of their criticisms.

Kitcher shows that one lesson had been learned – a more subtle theory of reference is required, and Kitcher advances a context-sensitive theory of reference, which he then applies within an interesting and detailed discussion of the phlogiston theory and its transition from Priestley to Lavoisier. Within the phlogiston theory 'dephlogisticated air' is of course a theoretical term and Kitcher suggests that such terms are not to be associated with just one single (putative) referent. Instead, each theoretical term has a 'reference potential' – a potential such that its tokens (i.e. utterances of it in various contexts) may refer to more than one (putative) entity, depending on what initiated that particular utterance, and that could be some event in which the referent is causally involved, or an event involving description. Kitcher (1993: p76) ties this reference potential to the Fregean notion of 'modes of reference' – since the reference of a term can be specified either by a description or in terms of the entity that *grounds* the utterance of this term, Kitcher suggests two modes of reference: a 'descriptive mode' and a 'baptismal mode'.

§6 Defending the Pessimistic Induction

Both Psillos and McLeish argue that there is no determinate way in which we can discover which mode of reference was in force for any given utterance, and they both refer to this as the 'decision problem'. Kitcher aims to solve the decision problem by employing the 'principle of humanity', which he attributes to Richard Grandy. That principle enjoins us to attribute to the speaker a

> pattern of relations among beliefs, desires and the world [which is] as similar to ours as possible.[10]

Kitcher then applies these ideas to Priestley's 'dephlogisticated air'. The idea is that, using the principle of humanity and what *we* know about Priestley's situation, we can reconstruct the situation and attribute to Priestley different dominant intentions on different occasions of his uttering the term 'dephlogisticated air'. More specifically, Grandy's principle enables us to single out some occasions on which our subject's behaviour can be best explained from our vantage point by attributing to him a dominant intention to refer to an entity that we now posit – oxygen. Take, for example, Priestley, when he used the phlogiston theory to characterise dephlogisticated air. For Kitcher, the right way to explain Priestley's using 'dephlogisticated air' on *those* occasions is by attributing to him the dominant intention to refer to 'what is left over when phlogiston gets absorbed by the calx of mercury'. Kitcher says that the mode of reference in those cases is given by this description

---

[10] See Grandy, 1973: p443. The principle is related to the *Principle of Charity* advocated by, for example, Davidson (1973) which counsels us to interpret speakers as if they held beliefs that are true (by our lights) wherever it is plausible to do so. The principle of humanity goes further, asking us to attribute to speakers beliefs similar to those *we* would hold if placed in similar circumstances. This again seems Whiggish.

§6 Defending the Pessimistic Induction

the substance obtained when the substance emitted in combustion is removed from the air (*ibid:* p102).

Since there is no such substance, such utterances of 'dephlogisticated air' fail to refer. But suppose we imagine Priestley on some other occasion where he has isolated a gas which he called dephlogisticated air, and he says of it that it supported combustion better than common air. Then, the right way to explain this utterance of 'dephlogisticated air' is that his dominant intention was to refer to whatever it is that actually supports combustion – i.e., oxygen. The mode of reference in those cases is given by Priestley's dominant intention 'to refer to the kind of stuff that was isolated in the experiments' (*ibid:* p102).

In an exactly similar manner, some of Fresnel's utterances of 'light wave' fail to refer because their reference is fixed by their theoretical description as oscillations of molecules of the ether, while the reference of others is fixed by Fresnel's dominant intention to talk about light, however it is in fact constituted, and these therefore refer to electromagnetic waves of high frequency. Kitcher concludes that claims of referential failure for past theories are overstated, and that past theories made many successfully referring (and true) claims about the world.

I shall in fact argue that Kitcher's proposed theory of reference does not do what he wants it to do. However, for the moment let's suppose that it did. Then he would have succeeded in showing that theoretical terms in discarded theories did *sometimes* refer, but only at the cost of running into both of those problems we encountered with Hardin and Rosenberg's proposals – the distrust problem and the ubiquity of reference problem. Taking the distrust problem first, Kitcher would have rescued the reference of central theoretical terms in discarded theories only when the

§6 Defending the Pessimistic Induction

scientist's usage of those terms clearly eschews accompanying descriptions, such as 'the substance emitted in combustion' or 'the oscillations of molecules of the ether'. Thus, for Kitcher, the realist has to accept that whilst some utterances of terms like 'dephlogisticated air' and 'light wave' in rejected theories did successfully refer after all, nevertheless the relevant theoretical descriptions of those entities were mistaken about virtually everything except the fact that the entities in question played some causal role in producing phenomena. But it is those theoretical descriptions which the realist hopes to defend in the case of current theories. Thus, Kitcher would have made no progress in defending realism from Laudan's historical challenge. For his account would show how some utterances of the central terms of past theories successfully referred in just those cases where their utterance did not depend upon those theories being right in their descriptions of the natural world. Kitcher's argument must again engender distrust in current theories; for if we are to assume that current theories bear some resemblance to past ones, then we should suspect that where they refer to theoretical items, those items either do not exist at all, or, if they do exist, our descriptions of them are probably quite false. This proposal should not be attractive to the realist, though it may be of considerable interest to many anti-realists, and possibly also to the structural realist, because it suggests that whilst our science can be successful in establishing that there are *some* items out there which have various structural relations to one another, we know nothing of what those items are like.

Turning to the ubiquity of reference problem, Kitcher's principle of humanity makes referential continuity too easily available, arguably ensuring that all abandoned theoretical terms can be made to refer to something. Hence, no abandoned concept has ever failed to characterise some item which our current theories now posit. It seems that, given the

§6 Defending the Pessimistic Induction

flexibility of the principle of humanity, we can always find explanations of why past scientists said the things they did such that *some* terms they used refer to items in our current theories. Part of the rationale for the judgement that Priestley referred to oxygen by some of his utterances of 'dephlogisticated air' was that on occasions he intended to refer to the stuff he had isolated, and this stuff was oxygen. But one can extend this line of thought to other abandoned expression-types posited in the history of science. Psillos illustrates the ubiquity of reference by pointing out that Aristotle could be said to be referring to a geodesic motion in curved spacetime when he talked about the natural motion of objects.[11]

I do not think Kitcher's critique of PI advances upon Hardin and Rosenberg even if his theory of reference was cogent. But I think it is not. For his analysis seems to render Priestley irrational in favour of maximising the coherence of *our* judgements of what he was doing, in the light of *our* knowledge and perception of his situation. As Psillos says:

> the principle of humanity does not offer a principled way to establish systematic correlations between Priestley's use of some tokens of 'dephlogisticated air' and intentions to refer to the stuff he had isolated which are not at the same time intentions to refer to this stuff as phlogiston-free air. (1997: p268)

In other words, by interpreting Priestley's utterances differently in different situations he is easily made to contradict himself. Consider these two simple examples that

---

[11] Robert Nola (1980) had already shown that a pure causal account of reference requires the addition of descriptive beliefs with his comparison of Thales, Gilbert, and Franklin as contenders for the discovery of electricity, with only the latter having the right descriptive beliefs to qualify.

§6 Defending the Pessimistic Induction

suggest Kitcher ends up making Priestley hold inconsistent beliefs.

Firstly, Priestley was clearly ready to assert both of these propositions:

p1   Dephlogisticated air improves combustion.

p2   Dephlogisticated air is produced when phlogiston is removed from ordinary air.

As far as the historical evidence goes, we may assume that he was also ready to infer from these that

p3   Dephlogisticated air improves combustion *and* is produced when phlogiston is removed from ordinary air.

Hence, the term 'dephlogisticated air' clearly was not ambiguous for Priestley, for whom, clearly, both tokens in p1 and p2 above refer to the same thing. p1 and p2 cannot be given different interpretations by us because we know Priestley would have affirmed p3 in which the same meaning is used throughout.

Secondly, the 'oxygen' hypothesis was certainly known to Priestley, and one could imagine him, being explicitly against that hypothesis, moving from the following two premises (p4 & p5) to the conclusion (p6):

p4   Dephlogisticated air exists.
p5   Oxygen does not exist.
p6   Dephlogisticated air exists but oxygen does not.

However, if the token of 'dephlogisticated air' in the first premise denoted oxygen, then Priestley would be contradicting himself. Kitcher's application of the principle of

§6 Defending the Pessimistic Induction

humanity seems to render some tokens referential, but only at the price of making Priestley appear inconsistent, if not downright irrational.

Both Psillos and McLeish devote considerable critical detail to discussions of Kitcher's proposals, and I leave the last word on Kitcher's proposed semantics to McLeish:

> [Kitcher's account] ... depends upon appealing to some fact or another which succeeds in distinguishing Priestley's referential tokens from those that failed to denote anything at all. In my view, no facts of this sort are available ... it is utterly uncontroversial that Priestley spoke some determinate truths that were important chemical insights, and that a good scientific realist semantics should be able to say how he was able to do so. Unfortunately, Kitcher's is not that semantics. (2005: p684)

To summarise, Kitcher's proposals suffer from similar problems to those of Hardin and Rosenberg. He proposes a semantic theory which recognises the inadequacy of a pure causal reference theory. This theory gains little general support and leads him into implicit attributions of inconsistency to Priestley. His attempt to divide the utterances of scientists into descriptive and causal reference types ends up making no advance as it still leaves the straightforward causal reference theory as central. Perhaps what is needed is an approach to reference which, rather than keeping separate descriptive reference and causal reference utterances, instead recognises that all reference involves a mixture of both descriptive and causal aspects. As we shall see, this is the cue for Psillos's proposals.

§6.3.4 Psillos: Causal Reference *and* Selected Descriptions

Kitcher learned lessons from Hardin and Rosenberg and offered a more sophisticated theory of reference, but nevertheless that theory was unable to form the basis for an adequate realist response to the PI. I will show that the story is similar in the case of Psillos. He learns lessons from Kitcher, being aware that realists do not gain by seeking to rely solely on referential continuity without descriptive accuracy for superseded theories. His own approach to a realist theory of reference (1999: ch.12) seeks to avoid this problem by demanding that at least some of the descriptive information associated with theoretical terms must be accurate before those terms can be said to successfully refer. Psillos rightly argues that pure causal theories of reference fail for theoretical terms because they make referential success too easy to attain. For there is always *some* item in the world that really causes the phenomena that led to the introduction of a theoretical term, so on a pure causal theory that theoretical term could hardly fail to refer – this is the ubiquity problem we have already encountered.

Because of this Psillos argues that any convincing account of reference will have to be causal-descriptivist – not only must a term refer to what caused the phenomena that led to its introduction into the language, but some of the important descriptions associated with the term must actually be satisfied by the item to which it refers. But he argues that not all the associated descriptions are equally important, and that

> some descriptions associated with a term are less fundamental in view of the fact that the posited entity would play its intended causal role even if they were not true. (*ibid:* p297)

§6 Defending the Pessimistic Induction

The important descriptions which must be satisfied by an item in order for a term to refer to it, are those comprising what Psillos calls the theory's 'core causal description' of the item concerned – the descriptions that must be true for the entity to play the causal role it has within the theory. Psillos is quite explicit here:

> the term which is employed to denote the posited entity is associated with a *core causal description* of the properties by virtue of which it plays its causal role *vis-à-vis* the set of phenomena. Insofar as these kind-constitutive properties comprise the causal origin of the core information associated with the term, then the term can be said to refer to this entity. (*ibid*: p295)

Thus, for example, the term 'phlogiston' failed to refer because

> there is nothing which fits a description which assigns to phlogiston the properties it requires in order to play its intended causal role in combustion. (*ibid*: p291)

Of course, this approach does allow the realist to choose which of the many item descriptions she will designate as the important ones, and she must do that from the viewpoint of how current theories describe the phenomena. This might be thought to be Whiggish[12] and *ad hoc*. We will come across this same concern in §6.6.7 where we will see Psillos again

---

[12] Some realists may say there is nothing wrong with taking a Whig approach to the history of science. My complaint is that it is implicitly (perhaps explicitly) question-begging in that it assumes that our present scientific theories are either true/approximately true, which amounts to RC1, or at least that they are more approximately true than theories of the past, which amounts to RC2. In addition, if theories of the past are evaluated in the light of current theories, it seems impossible for the RC2 convergence claim to be false.

§6 Defending the Pessimistic Induction

making judgements from our current perspective as to what was, and was not, important in previous theories. However, I will set aside this worry and allow Psillos and others to designate as important exactly those descriptions they want to associate with central terms in past theories so as to ensure they are referential in terms of current theories. Nevertheless, even with this worry set aside, we shall find that Psillos's arguments give him only another Pyrrhic victory. To see why, let us look more closely at the one case of reference Psillos examines in detail: the luminiferous ether of nineteenth-century optics and electromagnetism.

Considering the term 'ether' in the earlier wave theory of light, Psillos has to show that it did successfully refer because he acknowledges that it was a central term within that theory (with the descriptive implication that it served as a carrier for light waves). He accomplishes this by claiming that our modern term 'electromagnetic field' has the same causal description that 'ether' had in those earlier theories. Thus, Psillos says, the term 'ether' referred to the electromagnetic field itself. He also says that whilst current physical theory regards the beliefs of earlier 'ether' theorists about the *nature* of that ether to have been radically wrong, that doesn't matter because those beliefs weren't part of the core causal description of the term 'ether'. In other words, the false beliefs of earlier 'ether' theorists about the material nature of ether were not part of the causal description of the wave theory of light.

A problem here is that his case for the referential status of central terms in successful past theories invites a renewed form of the PI. For his proposal weakens our ability to distinguish, at the time a theory is held, which of our beliefs about the items referred to by that theory are part of its core causal description. For the historical record shows that those earlier ether theorists would strenuously disagree with

§6 Defending the Pessimistic Induction

Psillos's view of what it would be for an item to play the ether's causal role; that is, precisely with his claim about what is the ether's core causal description. Again, we shall see similar concerns in §6.6.7 as to Psillos's interpretation of history.

Stanford (2006: ch.6) examines in some detail what the ether theorists actually said, with a view to establishing what they would have thought the core causal description might be. For example, would Maxwell have believed that the ether could play the causal role ascribed to it in propagating light without it consisting of a material medium of some kind? Stanford is emphatic that he would not, and quotes the closing words of Maxwell's *A Treatise on Electricity and Magnetism* in support:

> If something is transmitted from one particle to another at a distance, what is the condition after it has left the one particle and before it has reached the other? If this something is the potential energy of two particles, as in Neumann's theory, how are we to conceive this energy as existing in a point of space, coinciding neither with the one particle nor with the other? In fact, whenever energy is transmitted from one body to another in time, there must be a medium or substance in which the energy exists after it leaves one body and before it reaches the other, ... Hence all these theories lead to the conception of a medium in which propagation takes place, and if we admit this medium as an hypothesis, I think it ought to occupy a prominent place in our investigations, and that we ought to endeavour to construct a mental representation of its action, and this has been my constant aim in this treatise. (Maxwell, 1955/1873: Vol. II, p493)

§6 Defending the Pessimistic Induction

Maxwell could hardly have been more explicit in asserting that only a material substance is capable of playing the causal role he assigns to the ether, and it is clear that this belief is of central importance for him. Thus, for Maxwell, the core causal description of the ether must include his beliefs about the ether's material character. But those are precisely what Psillos grants are not satisfied by the modern conception of the electromagnetic field (whether QED or its classical predecessor).

Even if Psillos can convincingly argue that, regardless of the beliefs of its theorists, the *actual* core causal description of the ether doesn't *need* to include anything about its material character, he would then find himself having driven a dangerous wedge between what the creators of scientific theories believe about their own theories, and what they ought to believe. In other words, the beliefs of leading proponents of successful theories as to which features figure in the core causal description have proved to be unreliable. Now the PI can be applied to those historical facts, leading us to conclude that we don't know which features of our *own* theories are rightly included in the core causal descriptions associated with *their* central terms, and that even the creators of those theories may be mistaken. So if Psillos is right then we have no way to pick out which parts of our best theories we can trust.

Thus Psillos achieves at most a Pyrrhic victory, because we are led to suspect that some parts of our current theories misinform us about the natural world, and are left unable to know which parts of those theories are correct. Psillos thus opens the way to considerable scepticism concerning our current theories, and that is clearly a threat to realism and a support to the kind of cautious scepticism which I advocate.

§6.3.5 Summary: Reference and Success

Realists have made many attempts to refute the PI by showing that previous theories which Laudan claimed were non-referring are such that their central terms actually did successfully refer. If this claim could be substantiated then this claim from §5.6 would be refuted:

LC1   There is no inference from the success of a theory to the fact that its central terms successfully refer.

Central to these attempts has been the use of theories of reference. Perhaps this has been a constructive research programme as successive proposals have been refuted by anti-realist arguments, leading to successive refinement of the concept of reference as it applies to theoretical terms. But common to all, including the most recent, has been the fact that they each gain only a Pyrrhic victory, for they each succeed in shortening Laudan's list (p121) at the expense of enabling a new PI argument to be run which concludes that our current theories cannot be trusted.

I also suggest that Bishop (2003) was correct to point out that nothing concerning the semantics of scientific theories can settle a debate as to whether the world is really as those theories describe it. Regardless of how the realist contrives his theory of reference to enable him to say that a term like 'ether' refers, there simply is no such elastic solid.

§6.4 The Mature Sciences Only Argument

I turn now to the type (ii) arguments (see p138) which involve the suggestion that if the realist restricts her claim for the link between success and approximate truth to just certain kinds of theories, rather than theories *tout court*, then all (or most) of the entries in Laudan's list can be removed.

One way realists have sought to achieve this is by saying that theories should not be counted unless they fall within a 'mature science'. As some realists themselves concede, that term is not clearly defined, [13] though its use is ubiquitous in the literature, right back to, for example, Putnam's 'Terms in a mature science typically refer' (1978: p20). Now it has to be admitted that the realist does have a point concerning an item on Laudan's list such as 'The Humoral Theory of Medicine'. It seems hard to think of that as a theory in the sense of other entries on his list such as the vibratory theory of heat, or the electromagnetic ether. Arguably it never was a theory but a long lasting tradition (from the classical period to the end of the nineteenth-century) which changed over time and whose boundaries were always vague. The so called 'Humoral theory of medicine' never helped Laudan's argument. However, that's just one entry – if the realist is to use this as a general argument then the phrase 'mature science' needs some definition. In §4 I argued that the phrase 'approximate truth' lacks precise definition. Here I will argue that 'mature sciences' is another term that the realist uses without much attempt at precise definition, nor, in this case, examination of the historical record to see whether the use of such a notion is justified. What the realist implies by the term is that for each domain of scientific enquiry, there is a point in time such that theories before that time do not conform to the higher ideals of scientific progress evidenced by theories coming after that time. Laudan himself quotes Krajewski: 'every branch of science crosses at some period the threshold of maturity.' (1977: p91)

---

[13] For example, Chakravartty: 'Maturity is an admittedly vague notion...' (2007: p8).

§6 Defending the Pessimistic Induction

While questioning the validity of the concept, Laudan suggests that the realist characterises a mature science as one in which

> correspondence or limiting case relations obtain invariably between any successive theories in the science once it has passed 'the threshold of maturity'. (1981: p34)

Thus Newton's theory could be said to be a limiting case of Einstein's. So a mature science could be said to have 'corresponding theories' – meaning theories in which these correspondence or limiting case relations obtain between themselves and their superseded predecessors. This view can be criticised on the grounds that the existence of such a threshold is neither confirmable nor falsifiable and has every appearance of being an *ad hoc* device aimed at limiting the impact of adverse historical evidence. It is not falsifiable because if it was discovered that no sciences yet possessed these 'corresponding theories', the realist could claim that eventually all sciences will. On the other hand, it is not confirmable because if we found a science whose successive theories were 'corresponding', we could not know if that relation would continue to apply subsequently. This latter point suggests that the realist just assumes that once a domain of science enters its mature phase it remains within it, but it is unclear why this should be assumed. Realists may be assuming that once a science enters this mature phase it is immune to any kind of Kuhnian revolution, with the special brand of problems relating to truth and reference that would accompany such an event. While such a view might be attractive to realism because it would give immunity to aspects of Kuhnian constructivism, it is surely implausible. Take the physics of the small as an example – it seems

§6 Defending the Pessimistic Induction

entirely credible that a complete revolution could occur there which would take the subject back into a 'pre-mature' phase.

A second difficulty for this concept of mature sciences, at least as a reply to Laudan's arguments, is this: by taking for granted that once a science becomes mature it will necessarily remain so, the convergent nature of scientific realism is simply assumed, and that is one thing which Laudan specifically sought to oppose, though this is an aspect of Laudan's work which I have not concentrated on.

A third difficulty is that many of the theories that were successful for a long time would be classed as falling outside the mature science group, but if realists wish to claim that science is a successful and progressive enterprise and that they have an explanation of that fact, then that explanation really ought to include the progressive successes that preceded these so called maturity thresholds. Claiming that the Crystalline Spheres or Ptolemaic theories of astronomy can be safely excluded from Laudan's list seems a dubious move for the mature sciences theorist to make. For while such a strategy would shorten Laudan's list, it could end up relegating much of the history of science to, in effect, not history of science at all, but anthropology, or history of some other subject, perhaps on a par with pre-scientific alchemy. Many of these immature theories enjoyed considerable predictive success and played an integral role in the history of their scientific domains, contributing to the emergence of successors, and, most relevant to my argument, enjoying great predictive success, yet the realist will have nothing to say about this. Any putative explanation of the success of science must include them, and yet they are excluded from the realist account of scientific progress. Realism seems unable to give an account of science that embraces its past as well as its present, a point we shall encounter again later (§9) when discussing explanations of the success of science, and the realist's inability to account for the success of false and superseded theories. Note that I have here deliberately picked upon two theories that would *obviously* not be classed as being mature sciences, but what of the vibratory theory of heat, or the optical ether – are these mature sciences? Is the line to be drawn simply at a point convenient to the realist argument?

My final objection is that this mature sciences argument is a proposal carrying too much of the whiff of Whig history, and risks becoming tautological. Boyd, for example, suggests that a science becomes mature when it reaches a point in its development

> at which the accepted background theories are sufficiently approximately true and comprehensive. (1981: p627)

But judgements of what is 'approximately true' can only be relative to our current understanding of what the truth is, so the criterion of maturity is just that this theory be sufficiently in agreement with our own current theories. Thus the only theories Laudan would be entitled to consider are those that already agree (approximately at least) with our current ones. With such a stipulation it seems that the claimed convergence of theories can hardly fail to be true.

§6.5 The Novel Predictions Argument

Here is another type (ii) argument aimed at restricting the claim of a link between success and approximate truth to just certain kinds of theories, rather than theories *tout court*, thus removing all (or most) of the entries in Laudan's list.

§6 Defending the Pessimistic Induction

Here the suggestion is that the only theories which should be regarded as 'successful' are those which make successful *novel* predictions. Alan Musgrave writes: 'The fact to be explained [by realism] is the novel predictive success of science' (1988: p. 239).[14]

One might point out that naturalist realists believe that their most powerful argument, the Explanationist Defence of Realism,[15] or NMA, should itself be regarded as a scientific theory, but I am not aware of it making any novel predictions, nor indeed any predictions of any kind.

In addition, it seems that realism is on the defensive here as its central claim becomes increasingly attenuated:

- The success of scientific theories is explained by their truth
- The success of scientific theories is explained by their approximate truth
- The success of mature scientific theories is explained by their approximate truth
- The novel successes of mature scientific theories are explained by their approximate truth

Controversy over whether novel predictions do confer additional epistemic virtue upon a theory goes back a long way, at least to the disagreements between William Whewell and J.S. Mill, who characterised Whewell's view as:

---

[14] Other philosophers taking this line are Lipton (1994), Psillos (1999), Sankey (2001), Ladyman & Ross et al (2007).

[15] See p105n6 and my discussion of the NMA in §8.

§6 Defending the Pessimistic Induction

> An hypothesis ... is entitled to a more favourable reception, if besides accounting for all the facts previously known, it has led to the anticipation and prediction of others which experience afterwards verified.
> (1843/1961: III, xiv, 6)

Mill is sceptical of this:

> Such predictions and their fulfilment are, indeed, well calculated to impress the uninformed. ... But it is strange that any considerable stress should be laid upon such a coincidence by persons of scientific attainments. (*ibid.*)

If the 'novel prediction' strategy is to succeed, and eliminate theories from Laudan's list, the concept of 'novelty' needs some precision. We might begin with *temporal* novelty – a prediction would be temporally novel if it was of something that had not previously been observed. However, this seems to introduce some arbitrariness into which theories should be believed. Surely whether someone had made this observation but didn't tell anyone about it shouldn't be relevant to deciding if it is novel or not. This is just what happened in the 'Fresnel white spot' case – it had actually been observed independently prior to its prediction by Fresnel's theory. So a temporal account of novelty may make the decision as to novelty just a historical accident.

It may be more plausible to base the criterion of novelty on whether some scientist knew about the result before the theory that predicts it was proposed, and we might call this *epistemic* novelty. This again risks making the criterion of novelty a historical accident, but in addition we now have the new problem that the scientist's knowing about a result may not undermine its novelty if she didn't appeal to this knowledge

§6 Defending the Pessimistic Induction

in developing the theory. For example, the success of general relativity in accounting for the previously anomalous orbit of Mercury is generally taken as highly confirming despite Einstein having been well aware of this anomalous orbit. This is because the theory was based upon general principles that weren't connected to empirical data about orbits of planets.

So now we move to what is generally called *use* novelty – a result is use-novel if the scientist didn't build the result into the theory or any of the auxiliary assumptions. Worrall (1994: p4) suggests realism is appropriate only for 'theories, designed with one set of data in mind, that have turned out to predict entirely unexpectedly some further general phenomenon'. However, Leplin (1997) rejects this as being too psychologistic. In the case of Fresnel, for example, if we say the fact of the white spot phenomenon being known about is irrelevant because Fresnel wasn't offering his theory to account for it, but it still predicted it, then we seem to be saying that scientists' intentions determine whether the theory's success counts as evidence of its truth. Leplin (1997) suggests that this undermines the objectivity of theory confirmation, and it does seem, in all of these cases, that we constantly end up saying that whether or not a theory should be believed is not determined by facts relating to the theory and its empirical predictions, but to accidents of history. This may seem particularly unattractive from the realist perspective.

Ladyman & Ross suggest a modal account of novelty – what matters for them is that 'a theory *could* predict some unknown phenomenon' and they also insist that what counts are only 'qualitatively new types of phenomena, which are then subsequently observed' (2007: p79). Presumably the thought here is that something like Newton's theory couldn't possibly predict the bending of light – it simply isn't within its scope to do so. However, this seems slightly problematic since it *might*

§6 Defending the Pessimistic Induction

turn out that Newton's theory *can* predict some phenomena which a later theory like Einstein's cannot, but we simply haven't yet encountered that phenomenon. The introduction of *could* seems to introduce too much vagueness. What exactly is meant by saying that a theory *could* make some prediction? Are we to assume that there are those phenomena which the theory *could* predict and has in fact done so, and those it *could* predict but which we know nothing of? How could we know that a theory *could* predict phenomena that we, as yet, know nothing of?

Then there is this insistence on wholly new types of phenomena being predicted in order to count as novel. On this view the prediction of the existence of Uranus by Newton's theory would not count as novel as it is not a new type of phenomenon. This begins to exclude a great deal – the return of Halley's comet would also presumably be excluded since returning comets are not new types of phenomena. I do not say that these issues cannot be resolved, but Ladyman & Ross have not done so in their brief discussion of this proposal.

The above is a brief discussion of some of the problems arising from the need for a precise definition of 'novelty' and it is not obvious that they can all be overcome. However, let's give realism the benefit of the doubt and assume that precision can be reached and that we do have such a useable concept.[16] The realist's proposal still faces problems.

---

[16] The modern literature on novel predictions is considerable, going back at least to Imre Lakatos (1970). A useful discussion of the history of the topic can be found in Leplin (1997: pp34-40). In his discussion of novel predictions Robert Carrier (1991: p26) refers back to William Whewell's 'consilience of inductions' (previously discussed) and Duhem's 'theoretical prediction of hitherto unknown laws'.

§6 Defending the Pessimistic Induction

These are examples of novel successes much favoured by realists:

- General Relativity: Light bending round massive objects, and the gravitational red-shift of spectral lines.

- Special Relativity: Time dilation (confirmed by jets carrying atomic clocks)

Realists seem to think it obvious that if a theory makes a novel prediction, then it *must* be true or approximately true. However, given the ability of an approximately true theory to make wildly wrong predictions (see §5.4), and the ability of completely false theories to make correct predictions, it is unclear why this intuition should be accepted. Moreover, as Mill believed, it is not obvious that the fact of some phenomenon being novel offers greater confirmation of a theory than is offered by its being non-novel, and thus Worrall reasonably asks:

> Was Einstein's theory better confirmed by the (novel) prediction of light-bending than it was by accounting for the already known facts about Mercury's perihelion? (Worrall, 1985: p34)

This is a genuine and valid question, and it is not at all obvious that the answer would favour the novel prediction, particularly if the question was addressed to scientists themselves. Worrall's question also highlights the uncertainty as to what qualifies as a novel prediction since, as discussed above, some might class the Mercury perihelion prediction itself as novel since that prediction was not part of the original motivation for Einstein's theory.

§6 Defending the Pessimistic Induction

I have yet to discuss the No Miracles Argument (NMA) which claims that if a theory makes successful predictions it must be true. The realist seems to be here appealing to a modified version of that argument which says that if a theory makes successful *novel* predictions then it must be true. As we shall see in §8.4 the NMA is an example of IBE and there is good reason to reject that mode of inference, and *if* that constitutes a reason for rejecting NMA then that rejection will still apply if theories are restricted to those having novel predictive success. Whether or not the predictions are novel, the move from predictive success to truth will still be an inference to best explanation.

Moreover, just as a reply to the NMA will ask how the successful predictions of false theories are to be explained, so I may ask for an explanation of the novel predictive successes of false theories. Realists seem to speak as though there are none, but there are many examples. Robert Carrier says that

> A viable anti-realist argument can only be based on cases in which wrong aspects of wrong theories are responsible for [novel] predictive success of these theories. (1991: p29)

He then proceeds to discuss two examples in great detail – Priestley's prediction of the reductive properties of hydrogen based on the phlogiston theory, and Dalton's and Gay-Lussac's prediction of the equality of thermal expansion of all gases based on the caloric theory of heat. (*ibid*: pp29-31) Carrier concludes that:

§6 Defending the Pessimistic Induction

> [novel] predictive success may well go along with lack of reference of the central terms employed. Reference is not necessary for [novel] success. This implies that the success of science cannot be even partially explained by assuming that the theoretical terms of successful theories are genuinely referential. (*ibid*: p32)

It has been my contention that 'lack of reference of the central terms' signals the absence of approximate truth (see p125), hence Carrier's work supports my contention that novel success does not imply approximate truth.

Timothy Lyons devotes three pages to no less than forty seven examples of the novel predictions of such false theories, and summarises them under these headings:

> caloric theory, phlogiston theory, W.J.M. Rankine's 19th Century vortex theory, Newtonian mechanics, Fermat's principle of least time, Fresnel's wave theory of light and theory of the optical ether, Maxwell's ether theory, Dalton's atomic theory, Kekulé's theory of the benzene molecule, Mendeleev's periodic law, Bohr's 1913 theory of the atom, Dirac's relativistic wave equation, and the original (pre-inflationary) big bang theory.[17]

Lyons also points out that 'Each of these theories has attained temporally novel predictive success. And, by present lights, not one of them can be true' (2003: p899). So realism cannot explain the novel success of these theories if truth is always the explanation of novel successes.

---

[17] This summary is given in Lyons (2003: p898) and the detailed list of 47 examples is in (2002: pp70-72) with 43 footnotes on pp87-89.

§6 Defending the Pessimistic Induction

However, even if it were granted that predictive novelty is of some significance, and thus that Laudan's list must be reduced to include just those theories that offer novel predictive success, further problems would remain for the realist. For it is not clear that this move would reduce Laudan's list to any important degree, since, as just mentioned, most of the important scientific theories of the past have made successful novel predictions. It seems that the realist will need to provide evidence that a significant number of theories on Laudan's list never made any novel predictions that turned out to be successful. Such evidence has not been offered, neither by Leplin (1997) nor Psillos (1999), nor anyone else to my knowledge.

Most superseded theories made novel predictions in their time – Newton's theory would be an obvious example. However, perhaps the most spectacular example would be the use of the nineteenth-century wave theory of light to make the novel prediction that there should be a bright spot of light at the centre of the shadow of a perfectly circular disc. Indeed, Poisson derived this prediction from Fresnel's formulation of the wave theory as an intended *reductio ad absurdum* which went dramatically wrong! Here is a novel prediction deriving from a theory which is now believed to be radically false, with its conception of light waves as oscillations of molecules of a material ether.[18]

---

[18] Note that the Fresnel white spot prediction would count as novel even by the more stringent demands of Ladyman & Ross. For it must surely count as a wholly new type of phenomenon. They concede that Fresnel's white spot was not temporally novel (2007: p77) but it was surely novel in *modal* terms, their preferred criterion. So it was modally novel and was a wholly new type of phenomenon, and yet it was nevertheless made by a theory now deemed false. Once again we see that whatever definition of novelty is used, false theories do make such predictions. Worrall (1994: p335) points out that, in addition, to the well known white spot case, Fresnel's theory of the wave surface inside biaxial crystals predicted the existence of internal and external refraction. Fresnel never even realised these results were predicted by his theory, the phenomena

There are many examples of successful novel predictions made by false theories, so the realist is no more entitled to infer truth or approximate truth from successful novel predictions than she is from any other successful predictions. Moreover, even if items can be removed from Laudan's list, those left, and there are many, remain susceptible to the PI argument – they remain as evidence that success is not confined to true theories. Since the historical record clearly shows that successful novel predictions can be, and often have been, made by incorrect theories, the continued belief that such predictions have a special link with truth is puzzling.

§6.6 Idle Wheels – Selective Truth Realism

§6.6.1 Introduction

Stated briefly, the realist believes that successful scientific theories are approximately true because success warrants a belief in approximate truth. However, the PI aims to sever the link between predictive success and approximate truth by showing that the historical record reveals numerous theories to have been successful which cannot have been approximately true.

In this section I shall examine the group (iii) arguments against this (see p138). We have seen the realist attempts to reduce the length of Laudan's list by restricting the kind of theory to 'mature' and the kind of success to 'novel'. Now we move to a different strategy in which the claim is that realism should never have claimed a link between success and the approximate truth of the *whole* theory, but only with the approximate truth of a part of it – that part that was actually

───────────────────── footnote continuation
themselves not even being thought of until after his death. This is novel prediction by any standard.

§6 Defending the Pessimistic Induction

responsible for the success, the non-essential part being referred to as 'idle wheels'. In effect the NMA is thus modified from the claim that a theory's success would be miraculous if it wasn't wholly approximately true, to the claim that the success would be miraculous if those constituents that had a bearing on its successful predictions weren't approximately true. For example, if the proposition 'the ether exists' had no actual involvement in the successful predictions made by theories in which the term 'ether' was used, then its being false would not count as the refutation of the NMA which the PI argues it to be. Consequently, all theories that mentioned it could be removed from Laudan's list (see p121). In this way the realist hopes that the historical argument against realism would be defeated.

In the literature a number of different phrases have been used to refer to this strategy – Hasok Chang (2003b) uses the phrase 'preservative realism', while M. Elsamahi (2005) refers to 'localised realism', Timothy Lyons (2006) to 'deployment realism', and Stanford (2003: ch.7) uses the phrase 'selective confirmation', deriving from the idea that evidence taken as confirming the whole theory in fact confirms only that subset of the theory which was responsible for its predictive success. I shall use the phrase 'selective truth realism' to describe this realist strategy because it attempts to describe as true or approximately true a selected part of an empirically successful theory, leaving the rest of the theory to be declared false. According to selective truth realism, a theory that includes entities or mechanisms that later turn out to be non-existent can still be regarded as approximately true if its predictive success did not depend on those false parts. If it can be shown that those parts of past theories that were responsible for their successes have invariably been preserved in subsequent theories then the same historical record which the PI attempts to use against realism would become a powerful argument *for* selective truth realism.

§6 Defending the Pessimistic Induction

This whole 'selective truth' approach might be said to flow from Hempel's view that confirmation cannot apply to all the statements of a theory, and that if it did then we could generate spurious confirmation for any statement, such as 'there are invisible fairies on the table', by simply adding it on to some well confirmed theory. And there are genuine cases of scientific claims or commitments which played no useful role in theories, such as Newton's commitment to the universe being at rest in absolute space. This is the well known 'tacking problem',[19] which is a serious concern for philosophy of science, this perhaps giving some reason to question the truth of 'confirmation holism' (see glossary).[20] This thought led Kitcher to say that 'confirmation does not accrue to irrelevant bits of doctrine that are not put to work delivering explanations or predictions.'[21] Kitcher goes on to propose that there are

> two kinds of posits introduced within scientific practice, *working posits* (the putative referents of terms that occur in problem-solving schemata) and *presuppositional posits* (those entities that apparently have to exist if the instances of the schemata are to be true) (1993: p149)

---

[19] Also known as the problem of irrelevant conjunction: if evidence E confirms theory T then it will also confirm theory T + C where C is any further claim that doesn't undermine E's confirmation of T. Thus if confirmation holism is unconditionally true then E *would* give spurious confirmation of C itself.

[20] Though not perhaps enough reason – see §6.6.4, p188.

[21] Kitcher invokes the support of Hempel (1965: ch.1) in criticising the holist approach to confirmation of theories: 'the hyperextended holism that spreads success over all accepted sentences without asking which have been put to work' and declaring that no realist 'should ever want to assert that the idle parts of an individual practice, past or present, are justified by the success of the whole' (1993: p142 & n21).

He later also refers to his presuppositional posits as 'idle wheels'.[22] For Kitcher, it is the working posits that are responsible for the success of theories, and they are also retained over theory change and can therefore be regarded as approximately true, thus establishing the link between success and approximate truth. When successful predictions are made we can thank the working posits and not the idle wheels. Kitcher then argues that so long as attention is concentrated on these important 'working' parts of a theory, the link between success and truth would be upheld despite the PI – in fact, subject to this re-interpretation, all entries in Laudan's list (p121) would be removed. Kitcher could then also claim convergence – that successive theories are improved approximations to the truth. So according to selective truth realism, when we say that some past theory was false, even though successful, all we are saying is that the *idle wheels* part was false, but the important *working* part was responsible for the success and was retained in the successor theory. Thus realists can claim that discarded theories included an approximately true part, while their embarrassingly false entities such as the caloric and the material ether can be ignored. I shall examine Kitcher's claims in more detail in §6.6.5.

Psillos also defends a position that amounts to the selective truth approach, giving it the portentous title of the '*Stratagema de Divide et Impera* Move' (1999: pp108-114). He wants to identify those constituents that are 'responsible for the empirical success of otherwise abandoned theories' (*ibid*: p108). He contends that realists 'need care only about those constituents which contribute to successes and which

---

[22] In, for example, his 2001a & 2001b. The term 'idle wheels' was perhaps a homage to James Clerk Maxwell. Peter Harman (1998: p5) describes Maxwell's ether model of vortices and *idle wheel* particles as 'the most famous image in nineteenth-century physics'.

can, therefore, be used to account for successes' (*ibid*: p110). I shall examine Psillos's claims in more detail in **Error! Reference source not found.** and §6.6.7.

Other philosophers who advocate this position are Leplin (1997), Niiniluoto (1999), and Sankey (2001). However, none of these has developed the selective truth realist's position in the detail that we find in Psillos. He proposes that our criterion for deciding on the *essential* posits is that 'Theoretical constituents which make essential contributions to successes are those that have an indispensable role in their generation' (*ibid*: p110).

This selective truth strategy again highlights the way in which realists seem to be gradually forced to dilute their claim. First we were told that theories are approximately true, with all the problems associated with that term, but now we are to think of theories as having only selected parts that are approximately true – perhaps such theories are selectively approximately true. Given the restrictions to mature theories and novel predictive success which I have already examined (§6.4, §6.5) the pressure on the claims of realism is evident. While it is an interesting attempt to defend scientific realism against the PI, the selective truth approach has some very questionable consequences and in the following I hope to show that it is a potentially disastrous strategy for realists to take as it fuels at least as much scepticism about our own theories as that which the PI threatens. I shall also argue that it simply isn't a convincing story since it creates a negative and unrealistic picture of the scientific enterprise that is not supported by the evidence of the historical record.

§6.6.2 The Notion of Idle Wheels is Problematic

It might be thought that this view of theories and their success creates a slightly odd picture; for it depicts a theory's idle wheels as doing no real work that contributes to its predictive

§6 Defending the Pessimistic Induction

success. However, it is clearly not plausible to suggest that those idle wheels are, therefore, completely redundant, since the very scientists who created the theory held them to be essential. This already suggests that they did *some* work since they seem to have been at least a psychological necessity for the scientists in order for them to have brought the theory into existence at all.

According to this way of thinking these idle wheels might be seen as only having a role within the context of discovery, in aiding the scientist(s) concerned to bring the theory into being. This seems implausible given the insistence on their indispensability made by those theory creators, evidence of which we shall encounter below. Perhaps it might be supposed that they constitute a kind of 'structural glue' that holds the theory together, and which assisted in the conceptualisation process by which the theory was brought into being, but which can later be dispensed with even though the scientists of that period didn't realise that to be the case. All this might be accepted, but what is remarkable in this proposal is that this structural glue turns out to be not simply dispensable, but wholly false and redundant when the context of justification is reached. Let us also remember that we are not talking here about items like Kekulé's snake – products of the imagination which were never believed to have any reality, and which were obviously only part of the psychology of discovery. Idle wheels are items which were postulated as real physical existents, and were regarded by the scientists as essential to their theory – like the material ether, or the substances known as caloric and phlogiston.

This results in a picture of science in which large numbers of theories, perhaps all, come into existence as a result of frameworks of assumptions which are simply false. Moreover, these assumptions are not willingly discarded by the creators of these theories as if part of the context of discovery, but, on the contrary, they are held onto and regarded as indispensable. The realist picture of theories convergent upon the truth seems to consist of one step backward and two forward, with scientists in a constant state of semi-delusion about their theories. I do not offer this as a decisive argument against selective truth realism, but it does seem to be a disturbing aspect of the proposal that it offers an unattractive view of the scientific enterprise. I believe it is sufficiently disturbing as to represent a *prima facie* motivation for thinking it wrong.

§6.6.3 Whig History?

Isn't the 'idle wheels' story perhaps a little too convenient for the realist? Being able to pick and choose from amongst the debris of our discarded theories would surely allow almost any thesis to be supported. Moreover, the picking and choosing is done from the perspective of our own current theories and their theoretical commitments. Given that fact it seems obviously possible to construct a successful convergence story. This is surely a Whig view of history of science, and if this is what the realist's view of the convergence of approximately true theories amounts to then it will be trivially true. For the selective truth realist asks two questions of any past successful theory – what parts of it were true, and what parts were responsible for its success? However, both of these questions are considered relative to our current theoretical beliefs about the world. These beliefs are used to decide both which parts of a previous theory are true, and which of that theory's parts made it successful. Given such an analysis it is unsurprising that Kitcher and Psillos find the same theory parts supply the answer to both these questions. Surely an obvious alternative explanation for this convergence is that both answers stem from a common source, namely our current theoretical beliefs, whatever those may be, and regardless of whether they are approximately true or not. Indeed, those current theoretical beliefs could even be

§6 Defending the Pessimistic Induction

completely false but would still yield the claim made by selective truth realism.

Of course this objection would lack force if realists could help themselves to the assumption that our present theories are in fact true (or approximately true, probably approximately true, partially true, or any of the many variations on this theme). However, to do that would beg the question. They may rightly argue that the truth (or approximate truth, etc.) of our current theories would explain the apparent agreement between the parts of past theories they claim were true and the parts they claim were responsible for the theory's success. However, the point here is that it would not then be the *truth* of our current theories that would be doing the explaining, for we have seen that they could equally well be false. Rather, it is the fact that the decision as to which parts of past theories are claimed to be true and which parts are claimed to be responsible for success, is decided based upon a common source, regardless of whether that source is true or false. What this makes clear is that in order to be successful the selective truth approach needs to provide objective criteria for how to identify the idle wheels of a theory – 'objective' here meaning independent of our current theories and their theoretical commitments. Without such objective criteria the approach accomplishes little as a defence against the historical record.

The problem for selective truth realism is made worse when we consider that history suggests that our judgments on this have been unreliable. Past scientists have repeatedly misidentified those parts of their theories which, according to realists, were required for their success. For example, we have already seen that J.C. Maxwell was emphatic that the wave theory required the existence of a material ether (see quote on p166), and I shall return to this issue in more detail below (§6.6.5). In addition, as we shall also see below, Lavoisier believed that a material substance was central to the caloric theory and its impressive success in explaining thermodynamic phenomena.[23]

Future generations may think that specific parts of *our* theories were idle and that the parts responsible for their success are preserved in whatever theories are then current. Nevertheless, today's realist cannot depend on our current judgments as to which parts those will turn out to be. Thus selective truth realism would represent a decidedly Pyrrhic victory for realism since it would leave us unable to reliably determine which parts of our current theories are required for their success, and which parts could be discarded. Without some reliable objective criterion for picking out the idle parts of theories, selective truth realism doesn't help the realist, for without such criteria we have no way to pick out the parts of our own theories which can be regarded as accurate, and the parts which are incorrect and misleading. Thus anti-realist scepticism concerning the theoretical commitments of current theories would remain justified.

§6.6.4 Theory Holism

The denial of confirmation holism is prompted by the tacking problem and the suspicion that theories may have idle wheels

---

[23] Stanford (2006, p167) considers how the selective truth approach would work if it had been applied in the past to theories then regarded as true. He considers what August Weismann might have thought of Charles Darwin's earlier false theory of the passing of heredity between generations. Weismann might have judged
> 'Darwin's commitment to the claim that individual hereditary particles are responsible for the developmental fate of particular individual cells to have been both something that Darwin got right and an important source of [Darwin's] theory's successes...' (*ibid.*)

But from our standpoint, both theorists were wrong. Weismann would have made this mistake because his theory supported the 'transmission of material units of heredity from parents to offspring' (*ibid.*) Thus judgements as to the *selective truth* of past theories are relative to the theory held by those doing the judging.

§6 Defending the Pessimistic Induction

that fall into the domain of that problem. However, the tacking problem does not mean we are *forced* to deny confirmation holism,[24] and indeed we would be entitled to simply ignore that problem on the grounds that it is not *our* problem at the moment. After all, it is a long standing problem in the theory of confirmation, and it is not obvious that we are obliged to solve it here in the midst of the realism/anti-realism debate.

In addition, the denial of confirmation holism is a very major step to take since it amounts to a rejection of the Duhem-Quine thesis and its applicability to scientific theories. If that step is taken then it would seem to open up many other issues and problems which I have not examined. However, what is more to the point, neither Kitcher nor Psillos have examined them either.

Regardless of the above, I think that realism itself is reliant upon theory holism. Laudan himself anticipated that realists might use the denial of confirmation holism to argue that parts of past theories were confirmed and approximately true even though those theories as a whole are now regarded as false. He urges that realists ought to be confirmation holists or else they will face a serious problem (1981, p28). Laudan assumes the realist believes that all the terms of theories successfully refer:

---

[24] It might be argued, for example, that the tacking problem derives from a distinctively *logical* view of confirmation and that a more *epistemic* view would dissolve the problem. See p224n27.

§6 Defending the Pessimistic Induction

R2 The observational and theoretical terms within the theories of a mature science genuinely refer (roughly, there are substances in the world that correspond to the ontologies presumed by our best theories). (1981, p28)[25]

He concedes that perhaps the realist only ever meant that *some* of the terms refer, quoting Putnam: 'terms in a mature science *typically* refer' (1978, p20, my emphasis). In other words, realists might argue that a theory's success warrants the claim that some of its central concepts refer, but not necessarily all. However, Laudan argues, this is a suspect move for realists since they have implicitly used confirmation holism to extend confirmational support from non-theoretical to theoretical parts of theories. After all, many anti-realists would deny that empirical evidence can confirm the theoretical parts of a theory other than via use of an inference such as IBE, which they reject. So whether evidential confirmation extends to the theoretical parts of theories goes to the very heart of the issue between realist and anti-realist.[26] If we ask why we should believe this we will typically be referred to this kind of claim by Boyd: 'experimental evidence for a theory is evidence for the truth of even its non-observational laws'[27]. This of course is not an argument but it at least makes clear the realist belief. So the realist relies upon confirmation holism to add legitimacy to her claims for the truth of the theoretical aspects of theories. Laudan says that

---

[25] This is different from the semantic claim which I attribute to realism in §2.3.5:
  RC3 Theoretical terms in scientific theories should be thought of as *putatively* referring expressions. Laudan's word 'genuinely' in his R2 makes his claim, in effect, an ontological claim rather than a semantic claim as to how scientific theories should be understood, a point made already on p124n10.

[26] Once again we see the theme that returns throughout this essay – the realist's need to establish a link between scientific method/predictive success and theoretical truth, first discussed in §3.7.4.

[27] See Boyd, 1973: p3. See also Sellars, 1962: p97

§6 Defending the Pessimistic Induction

if the tests to which we subject our theories only test portions of those theories, ... to be less than a holist about theory testing is to put at risk precisely that predilection for deep-structure claims which motivates much of the realist enterprise. (1981, p28)

Kitcher ignores this problem, but Psillos does address it (1999: pp125-6), claiming:

it is entirely consistent to stress that empirical evidence sends its support all the way up to the theoretical level, while recognizing that it does not do so indiscriminately and without differentiation. (1999, p126)

Psillos claims that remarks such as that of Boyd were never intended to support confirmation holism, but only the idea that empirical evidence could, in principle, supply confirmation for *some* theoretical claims. However, this doesn't seem to be an adequate reply to Laudan's anti-realist objection. For aside from assertions and plausibility claims, the only argument in support of Psillos's claim for selective confirmation lies in the historical record, and there, as we shall see, both Psillos and Kitcher fail to make their case. If he is to adequately reply to Laudan on this issue, Psillos needs to offer some additional argument, independent of the historical record, and this he fails to produce. Consequently the anti-realist (via spokesman Laudan) is in the position of already having been asked to accept without argument something he rejects – namely evidential confirmation of theoretical entities, and is now asked to accept the further stipulation that this confirmation can be selective.

Surely Laudan was right – what confirmation can there be of the theoretical items posited by a theory if confirmation holism is sacrificed? For no amount of observational evidence can, of itself, confirm statements about the theoretical other than by virtue of the part they play in the overall theory, but that means of confirmation is lost without holism. Support for the existence of a theory's theoretical items is considerably weakened if confirmation holism is discarded.

§6.6.5 Kitcher's Idle Wheels and the Material Ether

Kitcher does not give specific details as to how we can discern which parts of a theory are merely idle wheels. Instead he takes as an example the material ether which wave theorists believed to be the medium in which light waves and other electromagnetic phenomena are propagated:

> The ether is a prime example of a presuppositional posit, rarely employed in explanation or prediction, never subjected to empirical measurement ..., yet seemingly required to exist if the claims about electromagnetic and light waves were to be true. (1993: p149)

He claims that the successes of the wave theory didn't need the postulation of such a material medium, and that therefore those successes never did actually provide confirmation for it, and it can now be dispensed with. He says that for Fresnel

> ... the existence of such an ether was a presupposition of the successful schemata for treating interference, diffraction, and polarization, apparently forced upon wave theorists by their belief that any wave propagation requires a medium in which the wave propagates. All the successes of the schema can be preserved, even if the belief and the presupposition that it brings in its train are abandoned. (1993: p145)

§6 Defending the Pessimistic Induction

Here there is a Whiggish looking back via the understanding given us by our current theories – a re-interpretation of previous theories in such a way as to force them into a mould in which they can be made to appear to be approximations to our own. This can be seen again in Kitcher's reply to the quotation by Laudan (1981: p27) of Maxwell's famous belief that 'the aether was better confirmed than any other theoretical entity in natural philosophy':

> Maxwell was wrong. ... The success of the optical and electromagnetic schemata, employing the mathematical account of wave propagation ... gave scientists good reason for believing that electromagnetic waves were propagated according to Maxwell's equations. From that conclusion they could derive the existence of the ether – but only by supposing in every case that wave propagation requires a medium. (1993: p149)

If we are to deduce any generally applicable criterion that separates theoretical posits into working and idle, it seems that Kitcher suggests that the ether falls into the idle category simply because there was (or, more correctly, some philosophers now believe there was) an alternative possible theory which gives the same empirical predictions and explanatory story but without any embarrassing reference to a material ether. That theory would be that electromagnetic phenomena are propagated *like* a material wave, and in accordance with Maxwell's equations, but (somehow) without a transmission medium.

It is a questionable practice to, in effect, reconstruct previous scientific theories so that they will be in accord with a current philosophical doctrine. Moreover, what one might call the 'Pyrrhic victory syndrome' looms again, for this approach can be easily applied to most of those posits of contemporary

§6 Defending the Pessimistic Induction

theories that realists would claim to be genuinely confirmed by those theory's successes. An example would be one of Kitcher's own paradigmatic examples of a working posit, the gene, for we could suppose that characteristics are passed from parent to offspring in the patterns suggested by contemporary genetics, but (somehow) without the existence of genes themselves. Genes seem to be confirmed by their role in many explanations of the inheritance of organism characteristics, but only if, in each case, we presuppose that the inheritance of phenotypic traits demands some physical causal basis of the kind with which we are familiar. In other words, if Kitcher can produce an alternative theory such as 'transmitted like waves, but without any medium of transmission', rendering the ether an idle posit, then just about any theoretical posit could be given the same treatment.

Now in fact most current scientists would argue that if this was done to their theories the result would be unintelligible, and such would also be the case with scientists of the past. It was certainly the case with Maxwell and the ether, for he explicitly thought about whether the success of the wave theory could survive the jettisoning of the ether and rejected the very intelligibility of the idea of wave transmission without a material medium, as is clearly shown in the closing words of *A Treatise on Electricity and Magnetism*,[28] which demonstrates that Maxwell explicitly considered the alternative which Kitcher proposes – that the ether be considered an idle posit, and that he equally explicitly rejected it as incoherent.

An implication of Kitcher's proposal seems to be that if we suggest to current scientists that some of their theoretical assumptions are redundant, we should ignore their protestations of unintelligibility, reminding them of

---

[28] See the passage already quoted on p165 of this document.

Maxwell's fallibility, for (according to Kitcher) the record shows that scientists have continually proved to be poor judges on these matters. If Kitcher were right then the logic of the PI would work to show that the judgements of scientists couldn't be trusted.

Surely, we are not free to re-interpret an older theory in a manner that would render it unintelligible to those scientists who originally developed it. If that is what it takes to keep alive the story of that theory's success being due to the approximate truth of only a part of it then the realist seems to leave open the possibility that *any* theoretical posit of current science may turn out to have this idle status. This sows the seeds for a scepticism far deeper than that engendered by the PI which Kitcher seeks to undermine. I mentioned above the example of the gene, where we could suppose that characteristics are passed from parent to offspring just as contemporary genetics suggest, but (somehow) without the existence of genes themselves. Or perhaps light is transmitted as a stream of photons, but (somehow) without the existence of any actual photons. For if Kitcher's discussion of the ether is reasonable, then our current judgment of either of these examples as being incoherent is irrelevant to the fact that some scientist (or some realist philosopher) of the future may pronounce that it is the case and that genes and/or photons were, all along, idle posits.

Kitcher's view of what enables some posits to be categorised as idle seems extremely weak. However, perhaps one could be charitable and assume his intention was that a posit should be regarded as idle if it has no direct *causal* role within the theory, and that proposal might well work in the case of the material ether, which doesn't seem to have had such a causal role. However, this approach will not work in general as there are many discarded theoretical posits that did have a direct causal role in their respective theories – phlogiston and caloric fluid for example. Moreover, these discarded posits were the entities held to be directly responsible for producing the various effects which their respective theories predicted. So causal role considerations would not help Kitcher, for the successful predictions of past theories have frequently given causal roles to subsequently abandoned theoretical entities.

Kitcher's original motivation was associated with his Hempel-derived proposal that a successful account of confirmation must deny confirmation to completely unconnected parts of a successful theory that do no work in its predictions. Clearly he thinks that the problem of idle wheels, such as, he believes, the material ether, is analogous to the tacking problem. If a posit was just completely disconnected from the theory then it should perhaps be discardable, though it is not entirely clear that it is possible to prove conclusively that one of a theory's posits is 'disconnected', nor even precisely what that word means in this context.

However, what this historical case suggests is that the rejected posits of past theories, like ether, phlogiston, and caloric fluid, were no less intimately involved in the predictive successes of those theories than are genes, atoms, and the electromagnetic field in our current theories.

§6.6.6 Psillos and Scientists' Attitudes

Psillos's approach is broadly similar to that of Kitcher, in terms of claiming that selected parts of a theory are responsible for its success and that these may be carried over to later theories:

§6 Defending the Pessimistic Induction

> When a theory is abandoned, its theoretical constituents ... should not be rejected en bloc. ... If it turns out that the theoretical constituents that were responsible for the empirical success of otherwise abandoned theories are those that have been retained in our current scientific image, then a substantive version of scientific realism can still be defended. (1999: p108)

When discussing the reference of theoretical terms (§6.3.4) we saw Psillos's preparedness to select parts from previous theories based upon his modern perspective, and many of the points made there will recur.

We found that Kitcher's account of 'idle wheel' theoretical posits lacked a clear explanation of how the split into idle and working posits is to be made. Psillos attempts to rectify this by the suggestion that we need only look to what the scientists themselves said:

> My claim is that it is precisely those theoretical constituents which scientists themselves believed to contribute to the success of their theories (and hence to be supported by the evidence) that tend to get retained in theory change. Whereas, the constituents that do not 'carry over' tend to be those that scientists themselves considered too speculative and unsupported to be taken seriously. (1999, p112)

Any strategy relying upon our researches into the attitudes of past scientists sounds implausible and risky, and so it turns out to be. To begin with, it is ironic that we are advised by Psillos to rely on scientists' documented attitudes to their assumptions in order to identify the working posits. For while Psillos wants to rely upon the judgements of scientists in

§6 Defending the Pessimistic Induction

order to discern which are the working posits, as we have already seen in the discussion above on Kitcher's treatment of the material ether, the selective truth strategy itself undermines the reliability of the judgements of those same scientists concerning their theoretical posits. If scientists' judgements as to what is essential to their theories can be set aside, why should we then base our entire understanding of which are the working posits on what those same scientists appear to have believed?

However, there are more problems for Psillos's proposal. Analysing the attitudes of scientists is not a reliable procedure and is highly dependent on the contingencies of available historical evidence and its interpretation. And what about the ubiquitous disagreements in the attitudes of different scientists – how shall we judge them? Attempting to reconstruct the epistemic attitudes of past scientists involves dubious speculation about the extent of their commitment to each posit. Scientists may have epistemic attitudes such that they may not yet realise that a posit is actually indispensable, or alternatively, they may think it indispensable when it actually is not. Even if an entire scientific community sees a posit as dispensable that wouldn't matter if in fact the term really is indispensable to the theory's success. This begins to look like an exercise in historiography rather than philosophy and Michael Redhead rightly complains:

> the discussion looks not so much like philosophical analysis, but rather involves peering into the psychology and/or private notebooks to ascertain what scientists really meant by terms like 'ether' or 'phlogiston' (2001: p344).

In most instances there are no clues to their attitudes, but even if we did have access to authentic and unarguable evidence of scientists' attitudes towards their assumptions, why should we think these attitudes have much to do with the epistemic status of the theory's theoretical commitments? Psillos offers no reason for believing there to be a correlation between convictions of scientists toward their theoretical assumptions and their epistemic worth. Chang (2003b: p910) claims that the scientists which Psillos quotes as expressing doubt about caloric were in practice advocating caution towards *all* theories of heat. Stanford (2003, pp918-920) argues that many scientists strongly defended the reality of entities and mechanisms that turned out later to be non-existent. For example, right up to his death, Priestley defended the reality of phlogiston, and nineteenth-century physiologists maintained that vital forces were required in order to explain organic processes. As we have seen, nineteenth-century physicists judged the ether to be as well confirmed by the successes of the wave theory as it is possible for theoretical entities to be. That Maxwell misjudged (according to realists) the status of the ether so completely casts doubt on the view that we should look to the judgments of scientists to decide whether theoretical posits were of idle or working status. Kitcher and Psillos are here pulling in opposite directions – Kitcher saying the judgements of Maxwell and others were wrong, but Psillos advising us to rely upon those judgements.[29]

§6.6.7 Psillos and the Historical Record

Psillos (1999: pp115-145) discusses two examples which are supposed to exemplify his selective truth realism. These are the material ether, by way of an examination of nineteenth-century optics, and the caloric theory. I shall not here assess these in detail simply because I feel that, as examples

---

[29] Stanford (2006: pp173-180) provides an extensive refutation of Psillos's account of his selective truth realism, going into more depth and detail than I am able to here.

§6 Defending the Pessimistic Induction

supporting selective truth realism, they have been well and truly refuted in the literature, and it would be difficult to précis some very detailed historical arguments which are best examined in their original forms.

In the case of the material ether, most of the objections to Kitcher (above) also apply here. In addition, Stanford finds Psillos to have been selective in his reading of the historical data, and that his claims

> can only be defended by conveniently ignoring many such judgments made by leading scientists of the time, including the very scientists to whose more cautious and skeptical attitudes he appeals. (2006: pp174-5) [30]

Psillos also examines the caloric theory (1999: pp115-127) with a view to showing that the selective truth approach can remove this item from Laudan's list. His treatment of the caloric theory has been widely criticised on a number of counts, not least that his presentation of the historical data is too selective and partial. Stanford (2003, and 2006: pp175-9) criticises Psillos for presenting a biased reading of the history of caloric saying that the passages Psillos quotes are unrepresentative of their authors' attitudes, since they are drawn from isolated remarks made about caloric. Considering Psillos's (1999: p119) claim that 'scientists of this period were not committed to the truth of the hypothesis that the cause of heat was a material substance', Stanford complains that the textual evidence stands against this claim, referring to 'The strength of Lavoisier's confidence in the confirmation of

---

[30] Lyons (2006) is also critical of the Psillos analysis. Worrall (1994: pp336-7), though pre-dating the Psillos analysis, is quite emphatic that the material ether cannot be considered idle: 'The suggestion that the material ether was an idle component in Fresnel's system is significantly misleading'.

§6 Defending the Pessimistic Induction

the material theory of heat by the available evidence', and continuing:

> And once again, it seems that Psillos's case for the reliability (by current lights) of scientists' own judgments of selective confirmation must itself rely on an extremely selective reading of the historical record that ignores or dismisses many if not most of those very judgments. (*ibid*: p179)

In the context of an extensive discussion of the caloric theory, and referring to Stanford's earlier (2003) critique, Ioannis Votsis says that "Stanford's critical remarks about the bias in Psillos' textual evidence are reasonable by any standards" (2004: §5)

The most damaging critique of Psillos on the caloric theory comes from Hasok Chang (2003b), perhaps the foremost authority on the philosophical and historical aspects of the science of heat.[31] Contra Psillos, Chang argues that the laws of calorimetry cannot be seen as evidence for the preservation of the caloric theory of heat because they predated it, and are completely independent of it. He is also critical of Psillos's reading of the historical facts:

> We cannot take Psillos's version of the history, because there is little resemblance between what he describes as "the laws of the caloric theory" and the most important aspects of the actual caloric theories ... (2003b: p906)

---

[31] See, for example, his widely praised book on temperature (2003a).

§6 Defending the Pessimistic Induction

Chang also, in effect, accuses Psillos of doing Whig history, claiming that he

> has only picked out the items that have a clear appearance of being precursors of our modern beliefs. Even if this exercise were successful, it could only serve up a vacuous proof that those elements of past science that have survived into modern science have, indeed, survived. (*ibid.*)

Chang (*ibid*: pp906-910) mounts a convincing historically based argument for the complete indispensability of the caloric substance to the caloric theory. Following his survey of 'the most significant and undisputed successes of the most orthodox line of the caloric theory', he concludes that

> Almost all of them are neglected by Psillos; each of them tends to support the pessimistic induction, and fits Psillos's story only with great difficulty. The various assumptions about the nature of caloric and its interaction with ordinary matter did perform essential work in producing successful explanations, and they were clearly rejected by later science. (*ibid*: p909)

Psillos's historical evidence seems partial and at fault in the eyes of authorities who are more acquainted with the history than I am. I therefore do not take it as offering support for selective truth realism.

§6.6.8 Structural Realism

The widely influential position known as Structural Realism is now a major topic in its own right in the philosophy of science, and can be studied within the wider context of Structuralism more generally. van Fraassen has shown that the latter subject is not confined to the realist perspective, as he

§6 Defending the Pessimistic Induction

argues for 'Empiricist Structuralism' (2008: §11), whilst Otávio Bueno (1999) proposes 'Structural Empiricism'. Meanwhile, in addition to Worrall's 'Epistemic Structural Realism',[32] there is also considerable interest in a more metaphysical variant called 'Ontic Structural Realism'.[33] This is all a vast subject and it may appear ludicrous that I should devote such little space to it. However, as will become apparent, I need only consider the topic as it affects the Selective Truth approach to opposing the conclusions of the PI, and it is entirely from that perspective that I shall consider it here.

Worrall's Structural Realism ends up taking an approach similar to the idle wheels notion of Kitcher and Psillos, though in pursuit of a different aim. For unlike those others, Worrall accepts the force of the PI argument from the historical record, and sets out wanting to reconcile the NMA and the PI by proposing a way in which *both* these arguments can be accepted. To honour the NMA Worrall needs to establish some kind of link between approximate truth and theory success, and, like Kitcher, he tries to achieve this by associating that approximate truth not with the whole theory, but a part of it – in this case just that part concerned with *structure*. However, accepting the evidence of the historical record, he claims that continuity and cumulative progress between successive theories are nevertheless not lost because the mathematical structure of the superseded theory is retained in its successor. In addition, it is that mathematical structure aspect of the theory that gives rise to its success and

---

[32] The view that science can only ever inform us about the structural aspects of the world; that there may be facts of the matter beyond that structure, but science tells us nothing about them. See Worrall (1989, 1994).

[33] The view that scientific theories can only ever inform us about the structural aspects of the world, and there is nothing more to be known as the world is *all* structure. The literature is considerable, but see, for example: Ladyman (1998 & 2002), French & Ladyman (2003).

§6 Defending the Pessimistic Induction

which can be said to truly describe some structural aspect of the world. So Worrall's 'realism' again does not involve acceptance of every part of a successful theory, and his position is, in effect, similar to that of Kitcher and Psillos in that they both reject belief in the truth of a whole theory in favour of belief in the truth of selected parts.

The first point to make is that this hardly seems like a 'realism' at all, at least not in the sense I am discussing and opposing.[34] Yes it is a realism in the sense that it is committed to the view that something about scientific theories does manage to refer to something in the world. However, it is not that Worrall does not accept *all* the theoretical content of a successful theory, rather, he accepts *none* of it. Let me recall the two theses I take to be central to the realism which I wish to refute:

RC1 Most of the entities referred to by the theoretical terms of well-established current scientific theories exist mind-independently and have most of the properties attributed to them by science.

RC2 Scientific theories are typically approximately true and more recent theories are closer to the truth than older theories in the same domain.

Setting aside issues concerning 'approximate truth', clearly Worrall's strategy may enable RC2 to be defended, but only at the expense of RC1, which he surely concedes to be false. So in one sense Worrall's structural realism doesn't even enter the picture as seriously opposing the PI as it pertains to RC1. For what Worrall attempts is to find a way in which

---

[34] With its emphasis on the continuity of mathematical structure, Structural Realism can seem an attractive option in conceptual space for the anti-realist, but that is beyond my scope here.

continuity and convergence upon truth can be maintained, but what represents this continuity and convergence is simply the mathematical structure of theories, and not anything as specific as their actual theoretical commitments. For Worrall, all we can know of the nature of the world is its structure, on all else his epistemic pessimism seems similar to that of the anti-realist.

§6.6.9 Summary: Idle Wheels

The strategy of selective truth realism is to give up the holism of theory confirmation in order to enable the claim that previous successful but false theories can nevertheless be seen as containing a part that is approximately true, the very part that is responsible for the theory's success. This approach is based upon an analysis of the historical record that is partial, Whiggish, and highly contested by historical authorities. I summarise my other arguments against this strategy below.

§6.7 Conclusions

In this chapter I have examined a large number of realist attempts to undermine the PI. There is group (i) that attempts to show that Laudan is simply wrong in his assumptions, and thus fails to break the link between success and approximate truth and/or reference. Here I take myself to have shown the failure of successive attempts to establish a theory of the reference of theoretical terms which can sustain that link. I also argue that all attempts to use any kind of causal theory of reference beg the question against anti-realism – by assuming that theoretical terms do refer – but that is precisely what is at issue.

Then there is group (ii) in which realist claims are confined to only 'mature sciences' or those whose success yields 'novel predictions', this restriction being claimed to remove most of

§6 Defending the Pessimistic Induction

the entries in Laudan's list. However, like the term 'approximate truth' at the heart of realism, the term 'mature' fails to stand up under scrutiny. I offer various arguments against the significance of novel predictions, particularly when the historical record is taken into account.

Finally, in group (iii) I take myself to have shown that the selective truth realism strategy fails for two main reasons. Firstly, the interpretation of history given by its defenders is unconvincing, offering no objective way of deciding which are the true and success-conducive parts, and which are the 'idle wheels'. Secondly, it is a strategy that once again is such that, if successful, would represent a decidedly pyrrhic victory for realism since it would engender scepticism concerning our current theories, giving us grounds for expecting that some of their theoretical commitments will turn out to be false. All this at the expense of two major problems which the strategy would create – firstly, it fosters a view of scientists as invariably deluded about their own theories; secondly, by denying confirmation holism, realists undermine their own need for theory confirmation to extend to a theory's postulated theoretical terms.

The overall verdict must be that the PI has not been refuted and remains a significant problem for realism. If what science now says is correct, or approximately so, then the ontologies of previous theories are incorrect descriptions of the world even though those theories were successful. It follows that the success of our current best theories does not entail that they are correct in their descriptions of the world. We cannot infer approximate truth from theory success.

§7 Underdetermination and Unconceived Alternatives

§7.1 Introduction

We have seen that examination of the historical record gives grounds for scepticism about the claims of scientific realism. The next argument I shall present in support of the scepticism which I advocate is again based upon the historical record of science, though it is not normally thought of in that way. The problem known as the 'underdetermination of theory by evidence' (UTE) is a frequently recurring issue in the philosophy of science, but it is also of considerable importance in other areas of analytic philosophy. Underdetermination claims are often used to argue that our epistemic position with respect to some part of reality is not as good as we might have thought, and hence that scepticism is justified. Some well-known philosophical debates can be regarded as turning on whether or not a given underdetermination claim must be accepted, and thus whether or not we must adopt a more modest epistemic position concerning the issues concerned. In these other areas of philosophy sceptical claims can be seen as underdetermination claims. For example, radical Cartesian scepticism could be rendered as the claim that the evidence provided by my senses leaves underdetermined the two theories 'I am a person with an embodied brain' and 'I am a brain in a vat'. Similarly, scepticism as to the existence of other minds could be rendered as the claim that the evidence provided by observation of a person's behaviour leaves underdetermined the two theories 'The person has conscious experience' and 'The person does not have conscious experience but behaves as if she does'.

In this chapter I shall focus on underdetermination in the philosophy of science, and there one frequently encounters claims to the effect that a particular theory is underdetermined by the evidence, or even that all scientific theories are underdetermined by the evidence. In other words, even if we had access to all the possible evidence, we would still not be able to decide which of many alternative theories is true. If this claim can be upheld then it would threaten the general realist view that scientific theories reveal truths concerning 'hidden aspects of reality', and particularly claim RC1. For the sceptic can produce (or at least claim the existence of) theories that are empirically equivalent in that they say the same things about the observable phenomena, but make different and incompatible theoretical commitments. The sceptic can then claim that a realist view of such theories is not warranted since any observation that provides reason to believe one of the empirically equivalent theories gives equally good reason to believe the others, leaving only pragmatic reasons for theory choice.

Use of UTE as an argument against realism is often assumed to originate with Duhem.[1] However, Cardinal Bellarmine's attitude to Copernicus, often cited as the earliest occurrence of instrumentalism, might also be seen as an example of the UTE argument. Following the publication in 1543 of Copernicus's *De Revolutionibus Orbium Celestium*, some philosophers and astronomers took up the idea of a heliocentric universe with a moving Earth. However, when Bellarmine wrote to Foscarini about Galileo's work he suggested that

---

[1] See Duhem, 1914/1991: part 2 chapter 6.

§7 Underdetermination and Unconceived Alternatives

> ... it seems to me that Your Reverence and Galileo did prudently to content yourself with speaking hypothetically, and not absolutely, as I have always believed that Copernicus spoke. For to say that, assuming the earth moves and the sun stands still, all the appearances are saved better than with eccentrics and epicycles, is to speak well ...[2]

He was referring to the idea that the Copernican system was able to save the appearances – that Copernicus' theory could correctly predict the observations of the heavens. Whilst Bellarmine conceded that this theory might be considered better than the 'eccentrics and epicycles' of the Ptolemaic/Aristotelian system, in fact it used a similar number of such *ad hoc* devices. Clearly he judged that empirical observation couldn't enable a choice between the Copernican and the Ptolemaic/Aristotelian systems to be made and consequently that those two theories are underdetermined by the observational evidence.

Duhem took himself to be continuing/renewing Bellarmine's stance and wondered whether there might be alternatives to our best scientific theories that we had just never conceived of despite their being consistent with the evidence:

> Shall we ever dare to assert that no other hypothesis is imaginable? Light may be a swarm of projectiles, or it may be a vibratory motion whose waves are propagated in a medium; is it forbidden to be anything else at all? (1914/1991: pp189-190)

---

[2] Letter of April 12th, 1615. Reproduced from 'The Internet Modern History Sourcebook' at
http://www.fordham.edu/halsall/mod/1615bellarmine-letter.html

§7 Underdetermination and Unconceived Alternatives

He observed that in physics the testing of a particular theory or hypothesis always makes use of auxiliary assumptions,[3] and thus the two methods that might logically determine theory choice – falsification and verification – cannot do so since any failure to obtain the expected result can be dealt with by adjusting one or more of the auxiliary assumptions. If this is the case then pragmatic or other concerns must enter into the scientist's choice of which theory or hypothesis to accept. In other words, the strict observational evidence alone cannot determine which theory is accepted. But if those theories make incompatible references to theoretical items then anti-realism will be supported because those theories could not both be seen as true depictions of the way the world is.[4] Rather, each theory would be just one *empirically adequate* choice among many. Anti-realists, then, may appeal to Duhem's thesis in support of their interpretation of science, for it tends to undermine realist claims as to the truth of scientific theories and the reality of the objects of scientific investigation.

§7.2 The Standard Underdetermination of Theory by Evidence Argument

The standard UTE anti-realist argument has the following structure:

UTE1: For any theory there are other theories that are empirically equivalent – i.e. give the same empirical

---

[3] For example, initial and boundary conditions; principles, or other theories, concerning the measuring apparatus, etc.

[4] Karen Darling (2002) argues that Duhem's discussion of these issues was only intended to apply to theories of physics stated in mathematical form. If correct, this undermines the general applicability of UTE as stated by Duhem, but this is of historical interest only.

§7 Underdetermination and Unconceived Alternatives

predictions in all possible situations, but are incompatible in terms of their claims about the unobserved.

UTE2: If two theories have precisely the same observational consequences, then they are indistinguishable from an epistemic viewpoint, being both equally well supported by evidence.

UTE3: Hence there are no positive epistemic reasons to prefer one theory rather than the other and it follows that there is no reason to believe the truth of any theory.[5]

UTE1 is the thesis that all theories have 'empirical equivalents' – that is, theories that yield the same empirical predictions in all possible situations, and not just those that are presently known about. UTE2 is the thesis that observational evidence is the only epistemic constraint on theory choice, described by Douven (2008: p295) as the 'knowledge empiricism' claim.[6] Both are discussed below.

UTE has received considerable discussion in the literature,[7] and has been a fairly standard component of criticisms of

---

[5] It is instructive to compare the formulation of Igor Douven (2008: p294-5) which illustrates that he clearly assumes that the problem of UTE is exhausted by consideration of empirical equivalents:
(UTE1) EE: For each scientific theory there are empirically equivalent rivals.
(UTE2) KE: If the data alone does not suffice to determine a theory's truth-value, then nothing does.

[6] That observational evidence is the only *epistemic* constraint on theory choice, all other constraints being pragmatic in nature. Examples of the latter would be theory simplicity, fecundity, aesthetic appeal, consilience with other theories.

[7] Some examples are: Glymour 1971, 1980; Horwich 1982a, 1991; Kukla 1994; Laudan 1990a, 1990b; Laudan and Leplin 1991; Worrall 1982.

scientific realism.[8] Nevertheless, I think that, framed as above, it is inconclusive, and the debate that has taken place between realist and anti-realist over this particular argument has more or less reached a stalemate.

As we shall see, UTE2 is contested by the realist, and it could be said that the anti-realist's use of it is question-begging by simply taking for granted a verificationist stance which realists are unlikely to accept. UTE1 is even more controversial and in my view anti-realists have had long enough to come up with conclusive evidence supporting UTE1 but have failed to do so. In the following I shall argue that the underdetermination claims that have commonly been made in this field collapse into forms of more general radical scepticism.

However, I believe there is a version of UTE which has been largely overlooked, namely what Stanford calls the Problem of Unconceived Alternatives (PUA) and I shall argue that this represents a far more conclusive underdetermination argument against realism, albeit an argument of a very different form.

§7.2.1 UTE2: Other Grounds for Epistemic Acceptance?

I begin with UTE2 since it can be more speedily dealt with. Since it is the knowledge empiricism claim, it should be no surprise that realists reject it. UTE2 claims that the only epistemic grounds for acceptance of a theory are its observational consequences. What other grounds might there be? Some might claim that a theory is to be preferred to a rival if it offers some or all of the following advantages:

---

[8] For example, Duhem 1908/1969; Quine 1975. Many allege that van Fraassen's attack on realism relies on UTE (for example Kukla, 1998: p59; Worrall, 1984), but he strongly denies this (2007: pp346-349).

§7 Underdetermination and Unconceived Alternatives

(a) Simplicity
(b) Fecundity
(c) Aesthetic appeal
(d) Consilience with other theories[9]
(e) Explanatory value

It may be that possession of properties (a) – (d) would constitute reasons for accepting a theory as compared to a rival that lacks these properties. For example, given the nature of the scientific enterprise, perhaps the theory with these properties may be considered more likely to lead to a progressive research programme. Moreover, some realists would argue that (a) – (d) are truth conducive, and hence good epistemic grounds for theory acceptance, and hence they would reject UTE2.

But even if the realist concedes that (a) – (d) are pragmatic and non-epistemic grounds for theory acceptance, and thus do not entail theory truth, the anti-realist would still face a problem with (e). For there we encounter the principle of Inference to the best explanation (IBE), which claims that explanatory virtue can ground a belief in truth – if two theories both conform to the data, then one is entitled to believe true the one which better explains that data. Thus for realists (e) would constitute an epistemic ground for theory acceptance and they may then claim that UTE2 is a straightforward negation of IBE, and is thus question-begging against them.

---

[9] The phrase *consilience of inductions* was introduced by Whewell: 'The Consilience of Inductions takes place when an Induction, obtained from one class of facts, coincides with an Induction obtained from another different class.' (*The Philosophy of the Inductive Sciences*, 1840) The idea is that a theory is better confirmed if its predictions are also made by another independent theory.

Whether explanation does confer such an epistemic advantage will be discussed elsewhere (see §8.4). However, regardless of that discussion, there seems little point in considering UTE2 in the present context since it seems to boil down to whether or not one accepts IBE, and both sides to the argument will claim question-begging. As we shall see, IBE is a major point of contention between realists and anti-realists, and if UTE2 is important to UTE then there is a danger that the entire UTE debate will become just a repeat of the IBE debate. Thus UTE, considered as a good additional argument against realism, already seems to be in trouble since UTE2 is unlikely to be accepted by realists. Anti-realists are likely to accept UTE2, so for them the soundness of UTE will depend on UTE1, and to that I now turn.

§7.2.2 UTE1: Global Empirical Equivalents

How seriously should we take this speculative possibility that UTE1 alleges to be the case – that all theories have empirical equivalents? In the absence of evidence, why assume that such alternatives exist? It is understandable that critics of the UTE argument demand that examples be produced.[10]

UTE1 is a very sweeping claim concerning *all* scientific theories. So it is unsurprising that its supporters have sought to show that for any given scientific theory, there is always some kind of algorithmic means of generating (at least one) empirically equivalent theory. That is, an alternative theory that makes identical empirical predictions in all possible situations, and therefore cannot be distinguished from it by any possible evidence.

The approaches taken to the production of empirical equivalents can be divided into global and local varieties.

---

[10] See Kitcher (1993), Leplin (1997), Achinstein (2002).

§7 Underdetermination and Unconceived Alternatives

Global algorithms aim to produce empirical equivalents from *any* theory, T, and here are some examples that have been proposed:[11]

TE1 The theory stating that T's observable consequences are true, but T is false.

TE2 The theory stating that the world behaves according to T when observed, but some specific incompatible alternative otherwise.

TE3 The 'hypothesis of the Makers' – the (possibly incoherent) theory that we and our apparently T-governed world are part of an elaborate computer simulation.

TE4 The 'hypothesis of the Manipulators' – the theory that our experience is manipulated by powerful beings so as to make it appear that T is true.

One's initial reaction to these proposed theories may well be to think that they are not *real* scientific theories at all, but gerrymandered constructions aimed at supporting a shaky philosophical argument, and indeed, some have made this specific accusation.[12] Nonetheless, Kukla (1998, ch.5) attempts to defend these proposals.

However, perhaps this is all beside the point, for regardless of whether these are real theories, they surely amount to just another presentation of radical scepticism of the Cartesian 'evil demon' kind. Such philosophical fantasies pose an equal challenge to *all* knowledge claims, whether or not they are scientifically derived. Regardless of one's attitude toward the

---

[11] See Kukla (1993 & 1996), van Fraassen (1985).
[12] See Laudan and Leplin, 1991; Hoefer and Rosenberg, 1994.

§7 Underdetermination and Unconceived Alternatives

refutability of such scepticism, UTE was supposed to be a problem specific to the context of scientific theory, rather than just another example of radical scepticism. If such radical scepticism is the only reason we have for taking UTE seriously then there doesn't seem to be a UTE problem to worry about that is specific to the philosophy of science. Whether or not radical scepticism is a problem, it isn't *our* problem, here in this scientific realism/anti-realism debate.

There are other well known non-algorithmic attempts to demonstrate empirical equivalence to which we can make the same response. For example, the idea of a continuously shrinking or expanding universe, with concomitant changes in the dimensions of measuring devices making this undetectable. Although not algorithmic, it falls into the same category of radical scepticism. In a sense, these Cartesian fantasies merely amount to changing the subject by replacing our genuine concern about UTE in the scientific domain, with the different, but familiar, problem of radical scepticism.

Another proposal for a more or less algorithmic derivation of empirically equivalent theories involves taking the 'Craig transform'[13] of a theory as an empirically equivalent competitor. Even if we set aside the dubious plausibility of viewing a Craig transform as a valid scientific theory, we can quickly see that this idea will not do. The aim of the UTE argument is to show that claims concerning unobservable items are underdetermined – alternative claims are equally justified by the evidence – and that we thus lack warrant for regarding any such claims as *true*. The realist conceives of

---

[13] An influential theorem by William Craig implies that any scientific theory can be transformed into a form in which all theoretical terms have been eliminated, but which has all the observational consequences of the original theory. For a discussion of the history of Craig's proposal and the increasing rejection of its ability to play any role as an argument for instrumentalism, see Stanford, 2005: pp402-3.

scientific theories as in some sense revealing the 'hidden workings' of the natural world. However, even if the Craig transform could be considered to be a real, bona-fide theory, it would not offer an alternative account of those hidden workings. For the Craig transform approach simply says that whatever the evidence for belief in some particular theoretical account of the hidden workings, we are always free to trim our beliefs to only those aspects of the theory that are concerned with observable phenomena. However, such agnosticism concerning the hidden workings is defensible only if UTE (or some equivalent grounds for suspicion) is independently established, so it doesn't seem legitimate to cite Craig transforms as examples of empirical equivalence in an argument for UTE.

Perhaps it should be no surprise that the algorithms produced by philosophers cannot generate serious alternative scientific theories that are genuinely distinct from those that scientists have created. Genuine theory creation is an extremely difficult and demanding task, requiring the work of real scientists over many decades. The philosopher's search for an algorithmic procedure for generating empirically equivalent theories was always hubristic, and should now be seen as misguided and hopelessly ambitious.

§7.2.3 UTE1: Local Empirical Equivalents

In contrast to the global case, the local strategy tries to take advantage of some feature of particular theories to show that an indefinite number of serious empirically equivalent scientific theories can be derived by varying that one feature. The most obvious example arises in Newtonian Mechanics. Newton claimed that the universe is at rest in absolute space, and we may refer to that theory as $TN(0)$. However, his theory supports any number of empirical equivalents of the form $TN(v)$, where v ascribes some constant absolute velocity to the universe. However, such cases surely don't prove very much, for the realist will surely insist that this isn't a case of competing empirically equivalent theories, but just one theory conjoined with various alternative factual claims about the world for which that theory specifically entails that we cannot have any empirical evidence. Wise realists will confine their realism to those theoretical claims that can, at least in principle, have some contact with empirical investigation. Such realism should not extend to the conjunction of Newtonian theory with claims about the absolute velocity of the universe, any more than it should extend to the conjunction of Newtonian theory with any other arbitrary but unrelated claim, such as that the positions of the planets at the time of a man's birth affects his fortunes.

This example, and others like it, suggest that, just as in the global case, the local strategy again changes the subject from UTE to another long-standing philosophical problem in the theory of confirmation – the so called tacking problem (mentioned in §6.6). Like the radical scepticism encountered in the global case, this problem is philosophically serious but it can't be the sole reason for taking UTE seriously unless the problem of UTE is just equated with the tacking problem.

Nevertheless, some claim there are examples of empirical equivalents which are not subject to the above objections – they are not examples of radical scepticism nor trivial variations on a single theory that, in effect, reduce to the tacking problem. Perhaps the most convincing one is that proposed by John Earman, again referring to Newton's theory as TN:

> TN (*sans* absolute space) can be opposed by a theory which eschews gravitational force in favor of a non-flat affine connection and which predicts exactly the same particle orbits as TN for gravitationally interacting particles (1993: p31).

§7 Underdetermination and Unconceived Alternatives

This would represent a serious scientific theory and certainly isn't a trivial variation on TN. Earman mentions a few others (*ibid.*).

Another plausible case would be that of Special Relativity versus Lorentzian Mechanics, though this is more controversial since the latter involves the systematic expansion and contraction of all our measuring devices when in motion relative to absolute space, and that could be seen as an example of radical scepticism.[14]

Quantum Mechanics (QM) can also be seen as exemplifying a genuine case of empirical equivalence of theories. The several interpretations of the formal QM theory are empirically equivalent, but are, in effect, different theories which explain the world according to extremely different principles and mechanisms. For example, consider the earliest understanding of QM – the 'Copenhagen interpretation' – according to which a particle cannot simultaneously have a precisely known position and momentum. Contrast Bohm's 'Hidden Variable' interpretation, according to which particles always have a definite position and velocity, hence momentum; and on Bohm's theory, particles have two kinds of energy, the usual (classical) energy, and a 'quantum potential' energy. There are (at least) three very well-developed interpretations –

---

[14] Hendrik Lorentz proposed a way to resolve the problems raised by the 1880s experiments of Michelson & Morley on the motion of the Earth through the ether. He showed that if it is assumed that moving bodies contract in the direction of motion, then the observed Michelson-Morley effects would follow. This became known as the Lorentz–Fitzgerald contraction. Lorentz extended his idea, putting it on a firmer mathematical footing, and in 1904 published what became known as the Lorentz transformations. These transformations again figured prominently in Einstein's 1905 Special Relativity. Thus, it could be argued that the theories of Lorentz and Einstein are empirical equivalents, but are nevertheless quite different theories: for Lorentz retains Maxwell's ether, while Einstein proposes his relativity principles plus the assumption of the constancy of the speed of light. For a discussion of both theories see Stachel, 2000.

'Hidden Variable', 'Many Worlds', and 'Spontaneous Collapse'[15] that have the property that there is no observational way to tell them apart. Moreover, it seems that there *cannot* be an observational way to tell them apart. This seems like a very clear case of empirical equivalence.

It is interesting to note that all these more convincing examples of local empirical equivalence arise in the physical sciences, and that may suggest that there is something about the characteristics of physical theories which makes them susceptible to empirical equivalents. Laudan and Leplin (1991: p459) rightly point out that most claimed examples of empirical equivalence typically involve, in one way or another, the relativity of motion. This might lead us to suspect that there is something about such physical theories that renders them especially susceptible to the construction of empirical equivalents. It may even be that the ability to construct empirical equivalents is confined to theories within the physical sciences concerned with motion.[16] If we consider biology, for example, how would we construct a truly distinct and non-sceptical alternative to neo-Darwinism?

However, the most important thing to note is that none of these examples is generated by an algorithm that would have general application, and each is a very specific alternative to an existing theory. Surely such a paucity of convincing examples of local empirical equivalence doesn't provide much support for UTE1. If lots of serious examples of local empirical equivalence could be produced UTE1 would need to be taken seriously, but the difficulties, and rare success

---

[15] 'Many Worlds' is attributed to Hugh Everett. There are various spontaneous collapse theories, but the so called GRW theory of Ghirardi, Rimini, and Weber is an example.

[16] Realists who make this suggestion include S. Leeds (1994), Ernan McMullin (1991).

encountered, in developing even a few convincing examples of genuine empirical equivalence might seem to suggest the opposite conclusion – that the problem of empirical equivalents doesn't provide a serious argument against realism at all.

§7.2.4 UTE: Summary

Critics of the UTE argument are entitled to demand that serious examples of empirical equivalence should actually be produced, and Kitcher (1993) and Leplin (1997) have made this demand. If successful, the possibility of globally applicable algorithms that generate empirically equivalent theories would indeed be a powerful reply to critics such as these. However, I have argued that this approach collapses into a repeat of radical Cartesian type scepticism. In discussing UTE as an argument against scientific realism no progress will be made with arguments that are identical with arguments against global realism, and that is what radical Cartesian sceptical arguments amount to. The anti-realist should not expect to justify scepticism concerning the scientific realist's proposed knowledge claims by supporting radical scepticism of *all* knowledge claims. On the other hand, when we turn to examples of local empirically equivalent theories, these are very few in number and seem confined to the physical sciences, principally those concerning motion. Such a very small and localised sample offers no evidence to support UTE1, indeed the anti-realist is entitled to claim the opposite – that the paucity of evidence suggests that UTE1 is false.

Given that the argument between realist and anti-realist over UTE2 is, in effect, an argument about IBE, and thus a stalemate, that premise of the UTE argument will also not convince the anti-realist. Thus, when the problem of UTE is stated in terms of empirical equivalents, UTE2 ensures that it will not convince the anti-realist, and the lack of convincing evidence supporting UTE1 should ensure that it does not convince either realist or anti-realist. Clearly a fresh approach is required and to that I now turn.

§7.3 Unconceived Alternatives

It is worth pointing out that the search for empirical equivalents was just *one* way of showing that UTE is a real problem. It has in fact been the avenue that nearly all philosophers have pursued ever since Duhem raised the problem, and Quine highlighted it, and the literature is extensive. Unfortunately, the two issues – UTE and the problem of empirical equivalents – have become too firmly linked. Even the most recent major attack on UTE (Laudan & Leplin, 1991) and its most influential defence (Earman, 1993) both seem to assume that the problem of UTE is simply equivalent to the requirement to produce empirical equivalents. Igor Douven's article on underdetermination in a recent prestigious encyclopaedia (Douven, 2008) provides an example of the way in which it is simply assumed that the problem of underdetermination equates to the problem of discovering empirical equivalents.[17]

But the failure of 'empirical equivalents' to offer a serious UTE argument does not neutralise the seriousness of the threat against our current theories that underdetermination can pose. For that threat was not initially concerned with the possibility of empirical equivalents, but with the possibility of *any* alternatives sharing the empirical predictions of our best scientific theories. Our grounds for belief in a given scientific theory would be just as severely challenged if we believed there to be alternative theories that are not empirically

---

[17] See also p211n5.

§7 Underdetermination and Unconceived Alternatives

equivalent but which are consistent with the actual evidence we presently have in hand. The UTE argument has almost exclusively been that theories have empirical equivalents, and are thus underdetermined *in principle* and in all possible situations – that is, given all possible evidence. The approach I shall advocate claims that theories are underdetermined *in practice*, given the evidence known at the time of their adoption. I suggest that the real threat from the problem of underdetermination comes not from the sorts of philosophically inspired theoretical alternatives that we have considered, but instead from ordinary alternatives of the standard scientific kind that we have not yet conceived of.

Duhem's original statement of the problem did not assume that we needed to find empirically equivalent theories that are indistinguishable by any possible evidence. He simply assumed that there might be perfectly good alternative scientific hypotheses which had not even been imagined or thought of by us, but which would nonetheless be equally consistent with all of the *actual* evidence we knew of at that time. It will be helpful here to consider an example. Newton's theory and Einstein's Special Relativity are not empirical equivalents.[18] We know that the latter famously makes successful predictions that Newton failed to make. However, at the time Newton's theory was adopted, and for a considerable period thereafter, Einstein's was an unconceived alternative because all of the empirical evidence in support of Newton's theory which was then known about would in fact have been fully consistent with Einstein's theory if that had been conceived of. Of course it would be wildly implausible to suggest that Newton actually could have conceived of Einstein's theory at that time – one could imagine no

---

[18] I set aside Kuhnian incommensurability and assume that they do both refer to the same 'world' and are amenable to direct comparison in terms of their empirical predictions.

§7 Underdetermination and Unconceived Alternatives

justification available to Newton that would have supported any of the ideas central to Einstein's theory. Given Newton's location in history, any motivation for such ideas lay very much in the future. I shall return to this point in more detail in §7.4, but meanwhile, none of this is relevant to the assertion that at the time Newton proposed his theory there was at least one unconceived alternative. Hopefully, this example makes clear that it is not required that an unconceived alternative theory be even remotely like an empirical equivalent. For any given theory, the two questions – are there empirical equivalents, and are there unconceived alternatives – are quite different.

The phrase 'unconceived alternatives' derives from Stanford's recent work (2001, 2006), but Lawrence Sklar (1975) is one of the few philosophers to have referred to this phenomenon much earlier, though he called it 'transient underdetermination'. The idea is that the underdetermined theories are not empirically equivalent and therefore could be differentiated by the discovery of new evidence – their underdetermination is temporary or 'transient'. Given the considerable attention paid to the problem of empirical equivalents the importance of unconceived alternatives, or transient underdetermination, has been overlooked.

The question is how to decide whether there are unconceived competitors to *our* best scientific theories which would be equally consistent with all the *actual* evidence that is presently available to us. Sklar assumes that even those who are sceptical of empirical equivalence 'are likely to admit that transient underdetermination is a fact of epistemic life' (1975: p381). Sklar also says there are 'vast numbers of perfectly respectable scientific hypotheses ... we just haven't yet brought to mind,' including 'alternatives to our best present theories ... which would save the current data equally well' and probably even some theories 'more plausible than our

§7 Underdetermination and Unconceived Alternatives

own theories relative to present observational facts' (1981: pp18-19).

While it is obviously difficult to acquire convincing evidence regarding the likely existence of presently unconceived theories, as we shall see, the historical record of scientific inquiry provides evidence that 'unconceived alternatives' have been common. At times in the past, the scientific community has been unable to conceive of good alternatives to the then current theory. However, and this is a key point – the evidence that such alternative theories did exist is the fact that they were proposed and became accepted at a later date.

If unconceived alternatives can be shown to be ubiquitous in the history of science then, by simple induction, we should believe that our current successful theories also have unconceived alternatives, and that will undermine our justification for believing our present theories to be literally true. I propose that throughout the history of science, and in a wide variety of scientific fields, we have failed to conceive of other theories that were equally consistent with the evidence then available, and which, indeed, were subsequently adopted when new evidence became available. I call this the problem of unconceived alternatives – PUA.

In Stanford's work he refers to the 'New induction over the history of science' (NI). I arrive at an induction which is marginally different from Stanford's, and I refer to it as the 'PUA induction' (or just the PUA). In many situations these two terms can be used interchangeably. The PUA induction can be summarised thus:

PUA induction:

> Throughout the history of science and in nearly all fields of science, it has repeatedly been the case that only one theory was conceived of that was consistent with the

§7 Underdetermination and Unconceived Alternatives

available evidence and was well confirmed by it. However, radically different alternatives were later accepted that were equally consistent with that originally available evidence and which were, at that later date, regarded as well confirmed by all the evidence, including that which was available at the earlier date.

For example, if one considers theories of mechanics, as we move from Aristotelian to Cartesian to Newtonian to current theories, the evidence available when each earlier theory was accepted was equally consistent with its successor theory. Of course, the theory of relativity would never have been developed were it not for the anomalies that emerged in Newton's theory. Nevertheless, the account of gravitational motion offered by Einstein was equally consistent with the many phenomena for which Newton's theory had provided a convincing account. We find the same occurring in every area of scientific enquiry: we possess an existing theory that is well supported by some set of evidence. Anomalies come to light, and a new theory is proposed that was previously unconceived despite being equally consistent with that set of evidence that we had previously employed in support of the original theory.

Stanford (2006: pp19-20) offers the following list of examples to illustrate that this pattern occurs in a wide variety of scientific fields and historical situations:

Table 3: Examples of Unconceived Alternative Theories

- From elemental to early corpuscularian chemistry to Stahl's phlogiston theory to Lavoisier's oxygen chemistry to Dalton's atomic and contemporary chemistry.
- From various versions of preformationism to epigenetic theories of embryology.
- From the caloric theory of heat to later and ultimately contemporary thermodynamic theories.
- From effluvial theories of electricity and magnetism to theories of the electromagnetic ether and contemporary electromagnetism.
- From humoral imbalance to miasmatic to contagion and ultimately germ theories of disease.
- From eighteenth-century corpuscular theories of light to nineteenth-century wave theories to the contemporary quantum mechanical conception.
- From Darwin's pangenesis theory of inheritance to Weismann's germ-plasm theory to Mendelian and then contemporary molecular genetics.
- From Cuvier's theory of functionally integrated and necessarily static biological species and from Lamarck's autogenesis to Darwin's evolutionary theory.

In some of these cases it may be objected that changes in auxiliary hypotheses and/or background scientific beliefs were required before the alternative theories could be regarded as consistent with the originally available evidence. That is correct, but it is the implication of the PUA induction that in such cases the new auxiliary hypotheses will often be

§7 Underdetermination and Unconceived Alternatives

ones that were themselves unconceived despite being equally consistent with the available evidence. In such cases the total evidence available at the time of an earlier theory's acceptance is equally consistent with the combination of a later alternative theory together with the requisite later alternative auxiliary hypotheses.

It may be worth noting that the very concept of unconceived alternative theories requires the rejection of the most radical claims of incommensurability defended by Kuhn (1996). For according to Kuhn, the very phenomena themselves do not exist in any way that permits their identification across theories. In other words, the PUA induction requires that the constrained motion of a rock in a sling is one single phenomenon described differently by Aristotle and Newton, rather than Kuhn's view that Aristotle described a mixture of natural and violent motion while Newton described a completely distinct phenomenon of a pendulum losing energy through friction. However, rejection of these views of Kuhn should pose no problem in this context. To begin with, such radical incommensurability is among the least plausible features of Kuhn's account of science, and in addition, we are considering the PUA as an argument against realism and the scientific realist will probably not accept Kuhn's views on radical incommensurability anyway.

In support of his argument Stanford turns to the history of biology and gives a detailed case study on how Darwin, Galton and Weismann theorised about biological inheritance (2006: chapters 3–5). Each of these theorists conceived of inheritance in a specific way and did not consider important alternatives. Stanford's account offers considerable detail about how they approached inheritance, drawing on both primary and secondary historical sources. I shall attempt to summarise this.

§7 Underdetermination and Unconceived Alternatives

Darwin's theory of pangenesis proposed that development and inheritance were alternating links in a single causal chain.[19] The adult organism produces 'gemmules' and passes them on to the next generation. These gemmules are then seeds for development of a new organism. Darwin failed to conceive of any alternative to this causal chain view of development and inheritance.[20] However, Galton produced experimental evidence against the causal chain proposed by pangenesis with his transfusion experiments. By transfusing blood (and by hypothesis gemmules) into pure bred lines Galton tried to change the traits of offspring, but the transfusions had no effect. As a result, Galton proposed that development and inheritance were instead the products of a common cause, the underlying hereditary materials he called a 'stirp'.[21] Whilst holding different theories of inheritance, both Darwin and Galton shared a maturational view where hereditary materials act like seed crystals in a growth process.[22] It occurred to neither of them that the correct account might be that hereditary materials instead *direct* development.[23] Weismann later offered such a directive account that distinguished the materials that direct development (the germ) from the fully developed organism (the soma).[24] However, even Weismann failed to conceive of an important alternative. Like Darwin and Galton before him, Weismann had a view of inheritance in which hereditary materials exert their influence without exception. He thus failed to consider the more contextual

---

[19] *Ibid:* p68.
[20] *Ibid:* pp71–74, 81–84.
[21] *Ibid:* pp86–88.
[22] *Ibid:* p89.
[23] *Ibid:* pp89–94.
[24] *Ibid:* pp106–110.

alternative where the causal influence of hereditary materials depends on local cellular cues.[25]

So each of the three theorists failed to consider relevant alternatives to their views on inheritance, alternatives that became fully accepted by later scientific communities.[26] Stanford's case aims to show that the problem of unconceived alternatives occurred at every step of the progress of theorising about biological inheritance. This does seem to offer some inductive evidence from the history of science that the problem of unconceived alternatives is a very real one.

§7.4 Differences from Stanford's Account

Without doubt Stanford has placed the notion of unconceived alternatives on the map and my account follows his in many respects. However, there are two important points of difference. The first concerns Stanford's insistence that an unconceived alternative theory is equally well confirmed by the evidence available when the earlier theory was adopted. I think this is a mistake. Two different claims regarding unconceived alternatives are possible and I need to be clear as to which I am making.

(a) The unconceived alternative theory is equally well confirmed by the evidence.

(b) The unconceived alternative theory is equally consistent with the evidence.

---

[25] *Ibid:* p119.
[26] *Ibid:* pp132–133.

§7 Underdetermination and Unconceived Alternatives

Stanford's defence of his NI almost always uses the stronger claim (a). Many examples could be cited but here is one:

> Again the tough question, of course, is how to decide whether or not there really *are* typically unconceived competitors to our best scientific theories that are well confirmed by the body of actual evidence we have in hand. (2006: p18)

I think that this leaves him open to a serious objection; for in many cases, although the unconceived alternative theory is equally consistent with the evidence available at the time, it seems implausible to claim that it would be confirmed by that evidence. If we again consider Newton's theory, it is correct to claim that in 1687, when it was proposed, Einstein's later theory was equally consistent with the evidence then available to Newton. However, it is surely unhelpful to say that Einstein's theory would have been confirmed by that evidence. What evidence available in 1687 could reasonably have been taken as confirming the claim that the velocity of light is fixed, that the length of a body changes with increasing velocity, or that mass and energy are interchangeable?[27] I take it that a theory gains confirmation from some evidence if that evidence raises the likelihood of the theory being correct. However, no evidence available in 1687 would raise the probability of light having a fixed velocity, nor of the claim that the length of a body is dependent on its velocity. To some extent the likelihood of some claim is relative to the background assumptions and

---

[27] In this essay I favour an epistemic view of confirmation, though that is not defended here. I think that any theory of confirmation entailing the view that in 1687 Einstein's theory was confirmed by the evidence then available (and subject to the understanding of that evidence which was then possible), seems of dubious practical value to science.

§7 Underdetermination and Unconceived Alternatives

presuppositions of the period. Anyone in 1687 who proposed that the length of a body changes with velocity would probably have been regarded as slightly mad, or guilty of some kind of occultism, and all the evidence available then would not have seemed like confirmation.

I do not think that it was necessary for Stanford to formulate his argument using the stronger claim (a) since the argument of the PUA induction is equally strong using the weaker claim (b).[28] The essential qualifications for a theory to be an unconceived alternative are that it satisfies (b) and that it was later adopted by the scientific community as the best current theory at a later date. In other words, suppose that at date d_1 the earlier theory T_1 was adopted, and at the later date d_2, the theory T_2 replaced T_1. Then T_2 is an unconceived alternative for T_1 if T_2 is fully consistent with all the data known at d_1. There is no additional requirement that T_2 be confirmable at date d_1, only that it be confirmed at date d_2, but, by definition, that is the case since T_2 has been adopted as the best theory at that later date.

In the case where both theories are under consideration at the same period of history, then Stanford might be justified in making the stronger claim. For example, in the case that he examined, of Darwin and Galton, since these theorists were contemporaries working within the same background scientific culture, it may well be reasonable to say that not only was Galton's theory equally consistent with the data, but that it was also equally confirmed by it. However, when there is a large gap between d_1 and d_2 it becomes increasingly difficult to see the unconceived alternative theory as being equally confirmed, even though it is equally consistent.

---

[28] An alternative way of saying this is that Stanford did not *need* to rely upon a strict *logical* conception of confirmation, and that his argument remains valid if one adopts the weaker *epistemic* view that I favour.

§7 Underdetermination and Unconceived Alternatives

The second point of difference with Stanford's account is subtly related to the first. He portrays the problem of unconceived alternatives as concerned with failures of the cognitive capacities of scientists. They could have conceived of the unconceived alternative but just failed to do so. For example, he suggests that

> the history of science shows that we have repeatedly failed to conceive of (and therefore consider) alternatives to our best theories that were both well confirmed by the evidence available at the time and sufficiently plausible as to be later accepted by actual scientific communities. (2006: p29)

He later makes explicit his reference to limited cognitive capacities:

> While it seems possible to imagine cognitive supercreatures who are adept at conceiving of all possible theoretical explanations for a given set of phenomena ..., evidence suggests that we are simply not cognitive agents of this kind. (*ibid*: p46)

Stanford seems to assume that an unconceived alternative theory could have been conceived when the earlier theory was adopted if only the scientists concerned had possessed a more powerful cognitive capacity. However, this seems an incorrect assumption for reasons related to the earlier point concerning confirmability. Suppose that Newton had possessed unbounded cognitive capacity but had access only to the evidence available in 1687. Now let us imagine his mind cycling through every theoretically conceivable alternative theory. For each such theory he has to check whether it is consistent with the evidence available but also whether it seems likely – i.e. is well confirmed – given that evidence.

§7 Underdetermination and Unconceived Alternatives

When he came to Einstein's theory he would surely discard it on the grounds that it would not appear at all well confirmed, indeed, it would seem to be extremely unlikely given the evidence available in 1687. He would thus conclude that his theory was the right one to be adopted since Einstein's lacked confirmation.

I think Stanford is wrong to claim that the phenomenon of unconceived alternatives is necessarily a phenomenon concerning the cognitive capacities of scientists. This *may* indeed be so in some cases, but is not necessarily so. It does seem to be the case in the Darwin/Galton case researched by Stanford since these theorists were contemporaries working within the same background scientific culture. However, Newton's failure to conceive of Einstein's theory is no comment upon his cognitive capacities since many cultural and intellectual changes needed to occur before any scientist of Newton's period could have conceived of Einstein's theory.

I think that it was not necessary for Stanford to adopt the more powerful claim (a). For the PUA induction, based upon (b), is just as powerful even though it acknowledges that earlier scientists couldn't be expected to have conceived of the unconceived alternative. One might say that I am proposing an induction over theories that were unconceived and possibly *unconceivable* at the time of their adoption.

We have already seen that some see the PI as an induction over theories, though I do not rely upon this reading. Stanford uses this issue of failures in the cognitive capacities of scientists to ground his proposed contrast between the PI and his NI, claiming that the former is an induction over theories, while the latter is an induction over theorists and their inability to conceive of equally well confirmed theories:

§7 Underdetermination and Unconceived Alternatives

> the [NI] concerns the *theorists* rather than the *theories* of past and present science. That is, they point out not that past theories have ultimately been found to be false or otherwise wanting in some way ..., but instead that they were at one time *the best or only theories we could come up with,* notwithstanding the *availability* of equally well-confirmed and scientifically serious alternatives. (*ibid*: p44, Stanford's emphasis)

Of course, he is correct to say that his NI is different from the PI in that the former is not an induction over theories. However, the PUA induction can best be characterised as an induction over historical events. At each of these historical events a scientific theory was adopted by the scientific community, and we can now see in hindsight that a theory that was later adopted and replaced the earlier one, was in fact equally consistent with all the evidence that was known at the time of the event. Yes, in some cases the later theory would have been as well confirmed as the adopted one, and was not adopted for reasons of cognitive negligence, or any number of other reasons concerning the failings of all too human scientists. But only in *some* cases – in many other cases the most perfectly rational scientist could not have made a better choice of theory. Indeed, the most perfectly rational agent conceivable could not have chosen better, yet nevertheless, there still turned out to be an unconceived alternative theory that would later be adopted as being superior. This phenomenon flows from the inevitable historical rootedness of scientists who are cognitively embedded within a set of background assumptions and related theories and presuppositions which render the future theory virtually inconceivable. One is reminded here of Neurath's boat – the scientist cannot take the entirety of science into dry dock and rewrite all the rules, and that is what would sometimes be required in order for the future theory to be even conceivable, let alone confirmed. Instead the scientist is out at sea where he can make only relatively local changes to the ship that is the totality of scientific knowledge at that time.

A consequence of all this is that the conclusion of the PUA induction that I have proposed is different from that of Stanford's NI. Stanford concludes that we have probably failed to conceive of alternatives to our best theories which would be equally well confirmed by all the currently available evidence. However, the conclusion I draw from the PUA induction is that there probably are theories that would be equally consistent with all the currently available evidence. These theories have not been, and possibly *could not be*, conceived of by us, but will be recognised by our scientific successors to be equally consistent with all the currently available evidence, and will also be judged by them to be better confirmed given all the evidence that will then be available.

§7.5 Objections

In this discussion I refer to the three arguments – Stanford's New induction, the Pessimistic induction, and my PUA induction as NI, PI, PUA. The objections discussed here have been raised against Stanford's NI, and in general serve equally as objections against my PUA argument.

§7.5.1 The PUA relies upon the PI

A number of philosophers seem to think that the argument from unconceived alternatives is itself dependent upon PI. This may be because Stanford's entire discussion of Unconceived Alternatives frequently considers objections to PI as being potential objections to his NI; as a result of which there is a lot of space allocated in his book to the refutation of objections to PI. However, this is not done because NI relies upon PI, but rather because he thinks that several of the

§7 Underdetermination and Unconceived Alternatives

arguments against PI might also be levelled against NI. Devitt, for example, in a postscript to his (2004), whilst acknowledging the power of the NI seems to interpret it as merely a more powerful version of the PI, saying 'I think [Stanford's] version of the meta-induction is indeed the most powerful challenge.' (2010: p95).

Chakravartty sees the PI 'as a model for PUA' (2008: p152) and this influences the dialectic throughout his paper. However, as we shall see, I present the PI in a different way, not as an inductive argument at all, and presented in this way there is almost no connection between PUA and PI.

Psillos (2009: p70) is even more explicit, claiming that:

> The NI can only work in tandem with the PI. Unless PI is correct, NI does not suffice to show that the new and hitherto unconceived theories will be radically dissimilar to the superseded ones. Hence, rehabilitating PI is an important step in Stanford's strategy.

The reason for this claim can be found a little later when he makes the additional claim that the 'selective truth realism' arguments against the PI made by himself and others (discussed in §6.6) demonstrate that the non-idle parts of theories converge, and that this

> damages (at least partly) Stanford's unconceived alternatives gambit. If there is convergence in our scientific image of the world, the hitherto unconceived theories that will replace the current ones won't be the radical rivals they are portrayed to be. (*ibid.*)

So Psillos is claiming that even if NI were sound, we would have no reason to think that the unconceived alternatives

§7 Underdetermination and Unconceived Alternatives

would be *radically* different alternatives unless PI is also assumed to be sound. For Psillos believes that if PI can be defeated then he will be able to maintain his selective truth realism claim that some portion(s) of successive theories display convergence. It is hard to see why Psillos thinks this reliance of NI upon PI is so obviously true – he gives no additional justification for the claim, stating it baldly very near the beginning of his essay, and then taking it as justification for the fact that much of the rest of his essay consists of familiar objections to the PI which he has made elsewhere (and which are discussed and refuted in §6). This is ironic, as it demonstrates that, far from NI being reliant upon PI, it is Psillos's argument *against* NI which is completely reliant upon *his* arguments against PI.

I see no reason to believe Psillos's claim – Stanford discusses many examples of unconceived alternatives (see p227, Table 3) and some of these would certainly be described as radical. Perhaps Psillos has a more optimistic interpretation of what would be described as a radical alternative, and this is suggested by his later discussion of the move from Newtonian to Einsteinian gravitational theories where he claims that:

> the shift from Newtonian to Einsteinian is, by hindsight, not too radical ... the conceptual space of possibilities of Newton's overlaps with the conceptual space of general relativity. (*ibid*: p73)

I would suggest that the move from Newton's to Einstein's theory of gravitation exhibits radical conceptual change – the move from absolute to relative space alone justifies such a view, not to mention the latter's introduction of space-time. This illustrates that whether one thinks that the historical record displays convergence seems to depend upon one's point of view – one man's convergence is another man's radical dissimilarity. Ironically, if one takes an anti-realist stance and restricts both of these theories to their observable outcomes, then Psillos's claim here may seem reasonable, but it is hard to see how the theoretical commitments of these two theories can be said to be convergent and have an 'overlapping conceptual space of possibilities', and it is precisely with these theoretical commitments that realists must be concerned.

§7.5.2 The Mature Sciences Argument Again

Devitt (2010: pp96-97) correctly points out that a major realist line of argument against the PI has been to emphasise the immaturity of past sciences, thus justifying the claim that present theories are importantly different from those of the past. He concedes that this may be a somewhat *ad hoc* move, and then proclaims that the realist argument should always have been that our present scientific methodology is significantly better than it was in the past, and that this explains why present theories can be regarded as successful even though those in the past were not.

There are two major objections to this line of argument: To begin with, it is not clear why this is any different from the 'mature sciences' argument against PI already discussed (and refuted) in §6.4. However, while that was always an argument against the PI, Devitt offers no reason for why it should be viewed as an argument against NI or PUA. It is not obvious that improvements in scientific methodology have any bearing upon the overlooking of unconceived alternative theories. It has been a part of my variation on Stanford's argument that the reasons why theories were earlier unconceived but later come to be accepted includes the fact that, given the overall scientific context, they could not possibly have been conceived at that earlier period. The impossibility of Newton conceiving of Einstein's later theory has already been discussed, and even if the methodology at the time of Newton had been absolutely perfect, Einstein's theory would still not have been thought of. Secondly, Devitt overlooks the fact that even if it is conceded that current methodology is a considerable improvement over that of the past, at each stage in the past that has generally been the case. Thus we may expect that scientific methodology in the future will be improved yet further, thus leaving no reason to think that the fact that current scientific theories are a product of improved methodology gives any more reason for assuming them correct than it did in the past.

§7.5.3 Selective Truth Realism Again

Both Chakravartty (2008) and Psillos (2009) attempt to run the same argument against NI which we already encountered in §6.6 where I defended the PI against what I there called selective truth realism – the view that although previous theories have been discarded, there were aspects of those theories which were correct and were retained in the theories which replaced them. Chakravartty says that:

> The real question of interest here is whether there is anything like a principled continuity across scientific theories over time, which would allow realists to latch on to certain aspects of theoretical description as likely being approximately true. (2008: p153)

Psillos says that:

> recent realist responses to PI have aimed to show that there are ways to distinguish between the 'good' and the 'bad' parts of past abandoned theories and that the 'good' parts ... were retained in subsequent theories. This kind of response aims to show that there has been enough theoretical continuity in theory-change to warrant the realist claim that science is 'on the right track'. (*ibid*: p70)

In §6.6 I offered many reasons why this is not an effective realist strategy. Moreover, even if it was effective, it would still leave the realist with a Pyrrhic victory; for while Chakravartty, for example, says 'some aspects of current theories pertaining to unobservables are approximately true' (*ibid.*), we would then be in the position of not knowing which those successful aspects are. So the selective truth realism strategy here would lead to just as much of a sceptical problem concerning our current theories as the scepticism it was intended to dispel. Moreover, the entirety of this objection is anyway dependent on the assumption that NI depends upon PI, which it does not.

§7.5.4 Properties, not Entities

Chakravartty suggests that Stanford is overly concerned with entities at the expense of properties:

> Realism is too coarse if one conceives it at the level of entities, as Stanford and others do. ... Did all those theorists about electrons believe in the same entity? ... did they all believe in the property of negative charge? Yes they did. (2008: p155).

Aside from the fact that I see no evidence that Stanford does concentrate on entities and neglects properties, Chakravartty's argument is again liable to the same accusation of yielding a Pyrrhic victory, for if all previous scientists believed in electrons having the property of negative charge, but they got many other properties wrong, how are we to know which properties described by our current theories can be relied upon in the future? It is ironic that Chakravartty claims that a 'knowledge of unobservable properties and relations ... is no Pyrrhic victory for the realist' (*ibid.*) as Pyrrhic victory is precisely what it is.

§7.5.5 Psillos: the PUA is Self Defeating

Psillos claims that the PUA argument can be applied against Stanford himself:

> [Stanford] invites us to distrust the eliminative methods followed by scientists ... But hasn't Stanford undermined himself? Is he not in the very same predicament? ... Stanford must show that he is not subject to the kind of predicament that other theorists are supposed to be. (*ibid*: p74)

This objection is difficult to take seriously. Stanford proposed an induction over scientific theorists, and there is no reason to see any connection between those theorists and a philosopher. Moreover, if Psillos is to be taken seriously here, then not only has Stanford defeated himself, but Psillos may have defeated himself as well, for if there is a likelihood that Stanford has failed to conceive of some plausible alternative argument, then it is equally likely that so has Psillos – this way the ground opens up beneath us and there is nothing solid on which any of us can stand. Stanford argues that the historical record of scientific theory choice shows the frequent occurrence of failures to envisage alternative scientific theories which are later accepted as superior to the one which was actually chosen. However, this phenomenon arises from the complexity of scientific theories and (as I argue) the fact that the conceptual structures which they employ are relative to the totality of other scientific assumptions of their historical period. There is no justification to argue that if this PUA phenomenon is accepted then we cannot trust any of our beliefs about anything!

§7 Underdetermination and Unconceived Alternatives

§7.6 Conclusions

The conclusion of this chapter is that the attitude we should take toward the theoretical commitments of our current theories should include a degree of open minded scepticism. We should regard their theoretical proposals as good working hypotheses which may, and probably will, be superseded by theories whose theoretical import is quite different. Past scientific theories have been dogmatically championed, yet ultimately replaced by theories that wholly abandoned earlier assumptions. Moreover, objective examination of the historical record suggests that the existence of unconceived alternatives with different theoretical proposals is highly likely. Consequently, we have no grounds for regarding the theoretical proposals of our current theories as *true*, nor even *approximately true*. Thus RC1 is specifically undermined, and RC2 indirectly so.

§8 The No Miracles Argument for Scientific Realism

§8.1 The Argument Presented

Scientific realists claim that a notable success of science has been the giving of explanations of the world and its phenomena, and that inference to the best explanation (IBE) is the method used by science to accomplish this. When the very success of science itself becomes the phenomenon to be explained, we arrive at the No Miracles Argument (NMA) famously presented by Hilary Putnam as follows:

> The positive argument for realism is that it is the only philosophy that doesn't make the success of science a miracle. (1975, 73)

The claim is that scientific realism can *uniquely* explain the success of science. As we have seen, realists argue that scientific theories are approximately true and that most of the theoretical terms in such theories refer to items that exist, and this is just claim RC1. They then claim that the approximate truth of our theories is the explanation for why science works as well as it does, and, moreover, that it is the *best*, indeed the *only*, such explanation. Our theories are successful, they maintain, because they represent things as they really are. Many philosophers have presented the argument in this form.[1] James Cornman encapsulates this thought: 'what explains best describes best, and science explains best.'[2] Clearly he believes that theories which give good explanations will give

---

[1] Including Popper (1963), Smart (1968), Putnam (1975), Boyd (1983, 1984), Cartwright (1983), Fine (1984), Musgrave (1988), Leplin (1997), and Kitcher (2001a).

[2] Cornman (1976: p345) attributes this claim to Wilfrid Sellars (1962).

§8 The No Miracles Argument for Scientific Realism

predictive success, and I shall take Putnam's 'success' to mean predictive success, which is in line with my definition (p28n5).

The argument appealing to the success of science is not new, arguably being similar to the kind of argument made by Max Planck and Albert Einstein in the early twentieth-century debate over the reality of molecules. Even earlier, Clavius in the sixteenth century believed that the predictive success of Ptolemaic astronomy would be 'incredible' if that theory were not true:

> But by the assumption of Eccentric and Epicyclic spheres not only are all the appearances already known accounted for, but also future phenomena are predicted, ... it is incredible that we force the heavens to obey the figments of our own minds, and to move as we will, or in accordance with our principles (but we seem to force them, if the Eccentrics and Epicycles are figments, as our adversaries will have it).[3]

Of course, this example already suggests that all may not be right with the argument since we don't believe that Ptolemaic astronomy was true, so where did Clavius go wrong? He tried to prove too much, just as Galileo also did when he said, in a similar vein, that the earth must move because postulating that it does explains the tides.

Fine (1986a) points up an irony – that scientific realism hitches its own success to the success of science. Believing that 'the realist programme has degenerated by now to the point where it is quite beyond salvage' (p149), Fine comments that 'in so far as the successes of science mount

---

[3] From his *Commentary on the Sphere of Sacrobosco*, cited in Blake (1960: p34).

§8 The No Miracles Argument for Scientific Realism

while realism continues to decline we must conclude that scientific success lends no support to realism' (p150).

Though sceptical that the success of science needs any explaining, van Fraassen (1980: p39) calls the NMA the 'Ultimate Argument for Scientific Realism' and also uses Putnam's formulation. Boyd's (1983: p49) formulation is: 'If scientific theories weren't (approximately) true, it would be miraculous that they yield such accurate observational predictions.'[4] Stated thus it again suggests that the success of science has *only* one explanation, namely the (approximate) truth of its theories.

I have already referred to Kukla's formulation, which at least allows the possibility of other explanations:[5]

(1) The enterprise of science is enormously more [predictively] successful than can be accounted for by chance.

(2) The only (or best) explanation for this [predictive] success is the truth (or approximate truth) of scientific theories.

(3) Therefore we should be scientific realists.

However, I think this omits a premise. The argument is an example of IBE but needs an additional premise that infers truth from best explanation. The realist tends to be a naturalist and believes that what is good for science is good for

---

[4] Again confirming that it is *predictive* success that is referred to.
[5] See Kukla (1998: p12) which I criticised on p138. He agrees with Laudan's (1984b: p109) 'a theory is successful provided that it makes substantially correct predictions' again confirming that it is predictive success that we are concerned with. Henceforward if this is not made explicit, it can be taken as implicit.

philosophy.[6] She claims that science itself works by using IBE and we thus know that we can rely upon IBE to take us to the truth – IBE is a variety of inference that is claimed to be *truth-tropic*, to use Lipton's phrase (2004: p185). For realists the question of why one should be a scientific realist is to be treated in the same way as the question of why scientists accept some theory T. They believe that the answer to the latter is an explanation of the form

IBE_S: T is the best explanation of the relevant phenomena.

Hence, they claim, the realist answer to the question of why science is predictively successful should have the same form – namely, another inference to best explanation:

IBE_R: Theory truth or approximate truth is the best explanation of the phenomenon of scientific predictive success.

One may question the legitimacy of this parallel. When the scientist offers the explanation IBE_S it can be assumed that he has conducted various empirical tests, exposed the alternative theories to some experimental methodology, etc. However, nothing like the same can be said of the realist who offers explanation IBE_R. For testing IBE_R would require a stock of sample theories known to be both predictively successful and true/approximately true, but the realist can offer such a stock only by assuming what she wishes to prove – that current predictively successful theories are true or approximately true. And in the absence of such true predictively successful theories, the realist had better not refer to theories of the past

---

[6] For example, Psillos (1999: p73): 'the epistemology of science should employ no methods other than those used by the scientists themselves'; Boyd: 'The epistemology of empirical science is an empirical science' (1989: p13).

§8 The No Miracles Argument for Scientific Realism

for there she will find only the many theories that yielded predictive success without being either true or approximately true.

Furthermore, we can also question whether the kind of explanation which the NMA purports to offer is the kind of explanation which scientists value. Greg Frost-Arnold (2010) convincingly suggests that scientists only accept explanations which either generate predictions, or unify apparently disparate established claims. He then goes on to demonstrate that the NMA satisfies neither of those criteria and is therefore not the kind of explanation which scientists favour, and that, consequently, realists are not entitled to claim that the NMA resembles a scientific explanation. Frost-Arnold also points out that Worrall came to a similar conclusion much earlier when he pointed out the difference between a good scientific explanation, and the 'explanation' offered by the NMA:

> A requirement for a convincing scientific explanation is *independent* testability ... Yet in the case of realism's "explanation" of the success of our current theories there can of course be no question of any independent tests. (1989: p102, author's emphasis).

The realist's wish to ape the methods of science is a sham – scientific claims and philosophical claims are quite different and the methods of the former do not apply to the latter.

In any event, I shall use this as the formulation of NMA:

NM1   The enterprise of science is more predictively successful than can be accounted for by chance.

§8 The No Miracles Argument for Scientific Realism

NM2 The best explanation for this predictive success is the truth (or approximate truth) of scientific theories.

NM3 Since this is the best explanation, we should believe that it is true because science itself shows that inference from best explanation to truth is sound.

NM4 Therefore we should be scientific realists.

Note that NM2 claims the truth of scientific theories to be the *best* explanation, not the *only* one, as Putnam and Boyd claimed in their statement of NMA, a claim that is demonstrably too strong as other explanations have been proposed. I will later look at van Fraassen's alternative explanation, and there are also those of Laudan (1983), Carrier (1991), Stanford (2000), Lyons (2002: §6).[7]

Why the step from NM3 to NM4? Being a scientific realist seems here to be equated with believing in 'the truth (or approximate truth) of scientific theories'. Well, let me recall my preferred formulation of COSR:

RC1 Most of the entities referred to by the theoretical terms of well-established current scientific theories exist mind-independently and have most of the properties attributed to them by science.

RC2 Scientific theories are typically approximately true and more recent theories are closer to the truth than older theories in the same domain.

Setting aside the convergence claim included in RC2, it would certainly seem that anyone believing both of these claims

---

[7] Lyons calls his proposal *modest surrealism*, and it is based on Fine's proposal (1986a) which is itself based on Vaihinger's 'as if' fictionalism. Leplin (1997: p26) dubbed Fine's proposal *surrealism*.

must also believe in 'the truth (or approximate truth) of scientific theories'. So we could reasonably take the conclusion of the NMA to be that realism, as formulated in RC1, is correct.

The NMA is one of the most widely discussed arguments in the whole philosophy of science. In the following sections I will discuss some of its problems, and the many objections that have been raised by anti-realists together with their realist replies, and I will conclude that it is in fact fatally flawed on many different counts.[8]

§8.2 Approximate Truth is Ill Defined

The first problem faced by the NMA concerns its reliance upon the concept of approximate truth. I have discussed this at length in §4 and concluded that no useable definition of this phrase is available. Clearly, the NMA is reliant upon the concept of approximate truth, for if NM2 referred only to truth simpliciter, the entire argument would be obviously wrong since even realists would not want to claim that all current scientific theories are literally true. However, we have already seen in §4 that the phrase 'approximate truth' lacks definition and ensures that 'this whole debate is shot through with ambiguity'. Consequently, it is not clear that any formulation of the NMA can be stated in a manner that enables precise debate – it can't even get as far as being false. In the remainder of the discussion of NMA I shall simply have to speak as if a definition of approximate truth is available.

---

[8] I have relegated discussion of the 'Base Rate Fallacy' argument to Appendix 2 because although it tells in my favour and against the NMA, I do not think it is conclusive in the NMA/PI debate, indeed I think it misses the point.

§8.3 Approximate Truth Wouldn't Explain Success

I regard the PI argument as presented in §5 as the most conclusive argument against the NMA. I there presented the following four propositions as conclusions of the PI argument, and in §6 I examined the many objections to the PI and found none of them conclusive. I thus believe that these propositions are vindicated:

LC1 There is no inference from the success of a theory to the fact that its central terms successfully refer.

LC2 There is no inference from the success of a theory to the fact that it is approximately true.

LC3 The realist's claim that theoretical terms within the theories of a mature science genuinely refer lacks any argumentative support.

LC4 There is no evidence supporting an inference from the approximate truth of a theory to its success.

The key premise in NMA is:

NM2 The only (or best) explanation for this success is the truth (or approximate truth) of scientific theories.

But clearly LC2 and LC4 together show there is *no* link between success and approximate truth, and thus, if PI is valid, then, far from approximate truth being the best explanation for the success of science, it is no explanation at all.

I regard the combination of PI and the absence of clarity surrounding approximate truth as conclusive refutation of the NMA. Nevertheless, I will give further arguments against the NMA in what follows.

§8.4 IBE and Circularity

The next issue I shall consider is the accusation of circularity that has been raised in several quarters – that NMA begs the question against the anti-realist.[9] NMA is an example of inference to the best explanation (IBE), a principle according to which

> explanatory success provides grounds for belief in the truth of the successful explanatory story. (Fine, 1986a: p161)

But who is it designed to convince? It will have little force against those who reject IBE as a valid rule of inference, thus convincing only those who are already convinced. Laudan called the argument 'the Realist's ultimate Petitio Principii' (1981: p45). Fine says that the 'explanationist defence' of realism

> employ[s] the very type of argument whose cogency is the question under discussion. In this light the explanationist defense seems a paradigm case of begging the question, involving a circularity so small as to make its viciousness apparent. (1991: p82)

Kukla (a realist) also supports the circularity accusation and says

> the conclusion of [NMA] – that we have grounds for believing our scientific theories – doesn't follow unless its assumed that the explanatory virtues of hypotheses are reasons for believing them. (1998: p25)

---

[9] For example, Fine (1984: p86, 1986a: p161, 1986b, 1991: p82); Laudan (1981: pp45-6), Kukla (1998: pp24-26), Lipton (2004: ch.11).

§8 The No Miracles Argument for Scientific Realism

Another realist, Chakravartty, says 'As arguments go, the miracle argument is surprisingly poor' (2007: p5). In this section I want to show that this charge is valid, and also to raise doubts as to the general validity of IBE, at least in the form in which realists present it. Let me first recall the two central premises of NMA:

NM2   The only, or best, explanation for this success is the truth (or approximate truth) of scientific theories.

NM3   Since this is the only or best explanation, we should believe that it is true because science itself shows that inference from best explanation to truth is sound.

Examples of those who would reject IBE include Popper who rejects all forms of inductive inference (and IBE is a form of inductive inference), and van Fraassen who rejects IBE specifically.[10] Moreover, an instrumentalist will believe that only the observable consequences of a theory are true, and the realist wants the NMA to prove more than that. So it is indeed unclear who the NMA is intended to convince. NM3 is the key claim here, or rather the double claim:

(a)   Science uses inference from best explanation to truth.

(b)   The success of science demonstrates that such an inference method is sound.

But isn't (b) precisely what the NMA sets out to prove? As Lipton says, the NMA attempts to show that IBE

---

[10] See van Fraassen (1980: pp19-23 and 1989: pp131-150). Others who have also rejected IBE include Cartwright (1983), Laudan (1984: p28), Fine (1984: p5).

§8 The No Miracles Argument for Scientific Realism

> is truth-tropic by presupposing that [IBE] is truth-tropic, so it begs the question... The circularity objection does I think show that the miracle argument has no force for the non-realist. (2004: p186)[11]

At this point in my discussion I take the circularity of the NMA to have been demonstrated and I now move on to a more general discussion of IBE which will suggest further problems with it, including the claim to emulate the way in which IBE is used in science. This is claim NM3(a) and I shall present reasons for doubting it.

IBE is a method of argument that, in philosophical terms, has its roots in C.S. Peirce's *abduction*, but which is also used in everyday life as well as in science. van Fraassen gives a nice example:

> I hear scratching in the wall, the patter of little feet at midnight, my cheese disappears – and I infer that a mouse has come to live with me. Not merely that these apparent signs of mousely presence will continue, not merely that all the observable phenomena will be as if there is a mouse, but that there really is a mouse. (1980: pp19-20)

So the mouse hypothesis is taken as the best explanation of the phenomena consisting of scratching in the wall, patter of little feet, disappearing cheese. Note that van Fraassen justifies this use of IBE since it is an inference concerning the observable:

---

[11] At this point in Lipton's discussion he goes on to examine the issue of whether the NMA can be of use to those who are already realists. I will not pursue that avenue.

§8 The No Miracles Argument for Scientific Realism

For the mouse is an observable thing: therefore 'There is a mouse in the wainscoting' and 'All observable phenomena are as if there is a mouse in the wainscoting' are totally equivalent; each implies the other. (*ibid.*)[12]

But what kind of inference is involved in IBE? Which are the premises, and which is the conclusion? van Fraassen says 'I infer that a mouse has come to live with me', but this is not a deductive inference with the premises being scratching in the wall, disappearing cheese, etc. Generally, in IBE, the premises are the things to be explained (the *explanandum*) and the conclusion is the thing that does the explaining (the *explanans*). But in the standard deductive model of explanation we infer the *explanandum* from the *explanans*, not the other way around – we do not deduce the explanatory hypothesis from the phenomena, rather we deduce the phenomena from the explanatory hypothesis. Moreover, there is another reason why inferences of this kind cannot be of the deductive type, in which the truth of the premises guarantees the truth of the conclusion. For in IBE the conclusion, the *explanans*, goes beyond the premises, the *explanandum*; IBE is *ampliative*, deductive inference is not. IBE demands that, of a range of available possible explanations, we choose the *best* even though the actual best explanation may not be among those currently *available*. And that word 'best' introduces questions as to what criteria are to be used to make this apparently value based judgement. van Fraassen asks

---

[12] This also illustrates van Fraassen's allegiance to the old positivist view that empirically equivalent theories are indistinguishable.

§8 The No Miracles Argument for Scientific Realism

> The very phrase "inference to the best explanation" should wave a red flag for us. What is good, better, best? What values are slipped in here, under a common name, and where do they come from? (2002: p14)

We may recall (b) above which shows how the realist claims that philosophy should follow science in its presumed use of IBE because the fact that science uses it demonstrates its soundness as a method of inference. But this may lead philosophy into error. For even if it *was* agreed that science uses IBE to seek truth, it cannot be the *only* thing science produces – it clearly also yields practical benefits, interventions in the way the world goes and successful predictions. Consequently, if, in the course of producing these practical successes, science generated a few falsehoods, it perhaps would not matter too much alongside those practical triumphs. We have already seen how many previous theories were highly successful but false, so that a theory's falsity may not matter too much when viewed alongside its success. However, in the case of philosophical argument, only truth or falsity are on offer, so if IBE generates any falsehoods it may be disastrous for philosophy though of little consequence for science. Moreover, any use of IBE by science takes place against the accompaniment of an evolved and refined methodology of experiment which will tend to leave those false inferences of IBE lying discarded on the laboratory floor. Philosophy has no such compensating methodology. This is surely a prime example of the danger of assuming that philosophy should follow science in its method.

§8 The No Miracles Argument for Scientific Realism

Bearing in mind this cautionary note, I will now examine the structure of IBE:[13]

IBE1:

(a) F is a fact.
(b) Hypothesis H explains F.
(c) Of the available competing hypotheses that explain F, H is the best.
(d) Therefore, H is true.

IBE1 is not deductively valid but can be made so with the addition of premise (d) in the following:

IBE2:

(a) F is a fact.
(b) Hypothesis H explains F.
(c) Of the available competing hypotheses that explain F, H is the best.
(d) The best available explanation of a fact is true.
(e) Therefore, H is true.

However, this additional premise (d) is clearly false since the true explanation may not be included in the list of those under consideration. Moreover, falling back on approximate truth won't make it true, so the IBE2 formulation is not valid either. Can IBE be so obviously wrong? Musgrave may think so, as he claims that IBE 'thus construed is something that no sane philosopher should accept.' (1988: p238)

---

[13] In a manner that uses Musgrave's formulation (1988: pp238-9) as a starting point but ends up developing the argument against him.

However, there is a fallback position in which IBE may prove a weaker conclusion, but appear more plausible. This is by replacing premise IBE2d, a metaphysical proposition linking explanation and truth, with an epistemological premise, IBE3d as follows:

IBE3:

(a) F is a fact.
(b) Hypothesis H explains F.
(c) Of the available competing hypotheses that explain F, H is the best.
(d) It is reasonable to believe that the best available explanation of a fact is true.
(e) Therefore, it is reasonable to believe that H is true.

Already NMA, as stated, would be in difficulties as its claim is not that it is reasonable to believe that scientific theories are true, but that they *are* true. I also note that in the latest discussion of IBE by Psillos he concedes the need to weaken the claim in this way. Claiming that IBE cannot be seriously contested, he says:

> one cannot seriously question that the fact that a hypothesis stands out as the best explanation of the evidence offers defeasible reasons to warrantedly [*sic*] accept this hypothesis. (2009: p190)

Yet he overlooks this in his discussion of the NMA (*ibid:* pp49-52, 1999: pp78-81) where we find no mention of this weaker claim of having only 'defeasible reasons to warrantedly accept' scientific theories as true.

Returning to the discussion of IBE, more difficulties remain, for we need to understand what makes hypothesis H explain fact F, and also what it is for one hypothesis to be the *best* of a

§8 The No Miracles Argument for Scientific Realism

bunch of alternatives – as van Fraassen said (quoted above) of *best*, 'what values are slipped in here?' Some of these issues are discussed by Lipton[14] and one of particular concern is this: scientists may consider the best explanatory hypothesis to be that which is most elegant, or is the easiest to use, has the greatest coherence and fit with background knowledge, etc. It is unclear why factors such as this should be connected with truth, and here we encounter again the realist's problem, first discussed in §3.7.4, of needing to offer some connection between scientific method and theoretical truth. However, I want to raise a problem concerning that additional premise in IBE3:

IBE3d: It is reasonable to believe that the best available explanation of a fact is true.

This seems problematic, the difficulty being introduced by that word 'available'. Other explanations may in fact exist that we know nothing of, and one of those may actually be better, or even true! Lipton (2004: p152) calls this the 'argument from under-consideration', saying it is possible that 'the truth lies rather among those theories that nobody has considered.'

It is thus the case that, given a set of *available* alternative explanations, the *best* of them (however that is construed) may be far from the best of the larger set of *all possible* alternative explanations (some of which may be as yet unknown). van Fraassen (1989: p143) claims IBE is a rule which 'only selects the best among the historically given hypotheses. ... So our selection may well be the best of a bad lot.' Consequently, faced with a set of *available* alternative

---

[14] For example, whether *best* refers to *loveliest* or *likeliest*, discussed in his 2004: §4, §9, §11.

§8 The No Miracles Argument for Scientific Realism

explanations, the most epistemically favourable step may be to reject all of them and to suspend judgement until some better explanation becomes available. As Lipton (2004: p195) says 'the best must be good enough to be preferable to no explanation at all.' IBE, in effect, drops this latter option from consideration, as though, given any set of available explanations, it is *imperative* that one of them is chosen.

As an example, consider a religiously inclined person living prior to the discovery of evolution who believes the world is a few thousand years old. She is asked to explain the presence of fossils on mountain tops and seems to have these possible explanations to choose from:

a) God put them there.
b) Some creatures from another planet put them there.
c) We are deluded and they are just random markings in the rocks.

However, this way of presenting the choices open to this person omits the epistemically most favourable option – to suspend judgement, to say 'I do not know how they came to be there; it is a mystery awaiting a satisfactory explanation'. So in this case IBE3d is certainly not the best way to achieve a favourable epistemic outcome. We can accuse the realist of assuming that the set of *available* alternative explanations always includes what is actually the *best* explanation, thus overlooking the possibility that the best explanation hasn't yet been conceived of. This is, in effect, *the problem of unconceived alternatives*, discussed in §7. Clearly, for our pre-Darwin fossil hunter, there was indeed an *unconceived alternative* explanation – the theory of evolution combined with knowledge of geological sedimentation.

The realist might reply to this problem by requiring that explanations must be *satisfactory*, thus modifying IBE3 by the insistence that only *satisfactory* explanations qualify. Subject to that qualification, she would then regard IBE3 as valid, still not recognising the possibility that sometimes the best option may actually be to reject all the available *satisfactory* explanations. She would need to provide an account of when an explanation is satisfactory, but might well just hope for the best and assume that such an account can be given (just as Musgrave does in his 1988: p239). This seems very convenient for the realist who can simply announce that our best current scientific theories pass the test for being *satisfactory* explanations and that we are therefore justified in believing them true (or approximately so). But I do not believe that any impartial and objective account of *satisfactory* is available. Given the location in time of our fossil hunter, and the knowledge available to her, together with her background assumptions, it may well have seemed that (a) was a perfectly *satisfactory* explanation. Thus, according to our realist, she was justified in believing (a) to be true. However, to us (a) may not seem at all satisfactory. Similarly, it may have seemed to Maxwell that the best satisfactory explanation of electromagnetic phenomena involved an elastic ether, and thus he was justified in believing that theory true. Nevertheless, *we* may judge that theory to have not been satisfactory, and would certainly judge that it was not true; and our modern cosmologist maintains that a satisfactory theory makes her justified in believing it true that dark matter exists, but a century hence that theory may not seem satisfactory. However, if that is the way IBE is supposed to work then it seems to have a hopelessly psychologistic and context relative element at its heart.

The realist could reasonably dismiss an objection that the best available explanation might just be false, for any explanation

§8 The No Miracles Argument for Scientific Realism

might be false, but what we may *reasonably* believe isn't confined to certainties. Yes, it can sometimes be *reasonable* to believe what later turns out to be a falsehood. Even if van Fraassen's mouse explanation turns out to be false for some obscure reason, it was nevertheless reasonable for him to believe it – in such a case the weight of evidence pointed overwhelmingly towards the mouse hypothesis. However, this need not always be the case, and certainly isn't the case for our pre-Darwin fossil hunter.

Clearly there is a spectrum here and judgement is needed. If van Fraassen withheld judgement as to whether a mouse had come to live with him, just in case a better explanation came along later, we would consider him foolish, perhaps even epistemically irresponsible. Conversely, we might consider our pre-Darwin fossil hunter to be epistemically irresponsible for not suspending judgement. Most people will say that it was reasonable for Maxwell to believe that the best satisfactory explanation of the phenomena involved an elastic ether, but this is surely a dubious recommendation for the realist to make as it clearly shows that following IBE as a method of inference is not necessarily truth-tropic at all, as it certainly wasn't for Maxwell. Neither is it *approximately-truth-tropic* as our pre-Darwin fossil hunter shows that it is perfectly possible for IBE to lead to the inferring of an explanation that is neither true nor approximately true.

Where does this leave IBE3d? Here it is with the *satisfactory* qualification added:

IBE3d$_2$: It is reasonable to believe that the best available satisfactory explanation of a fact is true.

I think my discussion has shown that, as stated without qualification, this is not a safe premise on which to base a reliable general purpose epistemic policy. As van Fraassen's

§8 The No Miracles Argument for Scientific Realism

mouse shows, this does not mean that it is *never* safe to adopt IBE3d$_2$ and thus infer truth. However, it does suggest that IBE3d$_2$ should not form the basis of a generally reliable epistemic policy, and it thus seems unlikely that science does rely upon it. However, even if IBE is not central, science surely does include the use of some kind of ampliative inference such as IBE among its methods. So my proposal is that the realist makes an incorrect construal of the kind of IBE used within science. Something more like the following seems a better proposal for the way in which science employs IBE:

IBE4:

(a) F is a fact.
(b) Hypothesis H explains F.
(c) Of the available competing satisfactory hypotheses that explain F, H is the best.
(d) It is reasonable to use the best available satisfactory explanation of a fact as a provisional working hypothesis, always expecting, and being ready, to drop it if/when this hypothesis proves inadequate in the light of experiment, and/or better explanations of that fact become available.
(e) Therefore, it is reasonable to use H as a provisional working hypothesis, always expecting, and being ready, to drop it if/when this hypothesis proves inadequate in the light of experiment, and/or better explanations of that fact become available.

This is the correct, carefully sceptical, cautiously fallibilist approach to take, and note that IBE4 makes no mention of truth. We have already seen that scientists are indeed wise to adopt the strategy described by IBE4e above because the historical record shows the frequency with which they are unable to conceive of alternative theories that are equally consistent with the evidence and which are later actually adopted.

In the light of these remarks, what are we to make of NM3? It seems that NM3 is itself question-begging in favour of realism for it already assumes that realist presuppositions are at work at the heart of the scientific enterprise, because 'science itself shows that inference from best explanation to truth is sound'. However, that is so only when science is viewed from the realist perspective. The anti-realist can reasonably say that, viewed from her perspective, science shows no such thing, rather, it shows that taking the best explanation as a good, but fallible and provisional, working hypothesis is an epistemically successful policy, and that considerations of truth need not enter into it. As we shall see in more detail in §9, the anti-realist sees the methodology of science as the explanation of its success, rather than the realist's story concerning truth and successful reference.

The NMA is indeed circular but also attributes to science a truth-oriented version of IBE that it does not employ.

§8.5 Correspondence: Truth and Theory

In this section I shall examine the realist's use of the word 'truth', and the issue of *correspondence* in a general sense, not only as it relates to truth. My aim will be to show that scientific realism in general, and the NMA in particular, relies upon the notion of theory-world correspondence, and that this itself is reliant upon IBE.

The realist position on truth, as captured in NM2, relies upon some concept of *correspondence* – for the realist, theories describe a mind independent reality, and the theoretical terms of those theories *refer* to items that exist as part of that reality. Now in fact, as we have previously seen, most realists take

§8 The No Miracles Argument for Scientific Realism

this further, explicitly assume the correspondence theory of truth, and oppose epistemic theories of truth such as, for example, pragmatist or coherentist. However, I wish to argue that even if they do not, they must be committed to a notion of theory-world correspondence which leaves them open to considerable problems.

Not all realists subscribe to the correspondence theory of truth, and Devitt denies any link between realism and truth at all, claiming that 'no doctrine of truth is in any way constitutive of realism.'[15] Nevertheless, many realists do subscribe to the correspondence theory of truth, and some concede the problems which this gives them. For example, Musgrave (1996: p24) observes that 'The most fundamental objection to scientific realism concerns the realist conception of truth.' Table 1, p30 shows several realists who subscribe to the correspondence theory of truth, and others would agree to its central importance. For example, Putnam (1978: p18) claims that 'Whatever else realists say, they typically say that they believe in a correspondence theory of truth.'

Kitcher (1993) includes a section titled 'Rehabilitating Truth' (pp128-133) in which *truth* is more or less equated with *correspondence truth* and realism with acceptance of that view of truth, which is regarded as simple common sense:

> The correspondence theory of truth is often held to involve extravagant metaphysics, but, I claim, its roots lie in our everyday practices. (p130).

Niiniluoto includes among his basic tenets of realism:

---

[15] See Devitt (1997: pp43). However, it should be noted that, although he claims that the thesis of realism is not related to 'truth', he does in fact advocate a correspondence theory of truth, see, for example, his (1997: pviii).

§8 The No Miracles Argument for Scientific Realism

> Truth is a semantical relation between language and reality. Its meaning is given by a modern (Tarskian) version of the correspondence theory... (1999: p10)

Chakravartty confirms that many realists

> are uncomfortable with epistemic theories of truth, and adopt instead some version of the correspondence theory, according to which truth is some sort of correspondence between things like theories and the world. (2007: p13),

Howard Sankey includes the correspondence theory of truth as one of his six central constituents of scientific realism, claiming that

> truth consists in correspondence between a claim about the world and the way the world is. (2008: p16)

Despite his reservations mentioned above, Musgrave says that

> The aim of science, realists tell us, is to have true theories about the world, where 'true' is understood in the classical correspondence sense. (1988: p229)

The notion of correspondence raises well known issues that spread beyond the boundaries of the merely *scientific* realism/anti-realism debate and saddles the realist with classical objections to the obscurity of the correspondence relation. The most well known of these is what I term the *independent access* problem – a theory makes statements about the world, but how can we judge the truth, *qua* correspondence, of what it says about the theoretical, as we have nothing to compare against? In the paper in which Ellis attacks the reliance of most realists on correspondence truth and advocates his internal realist view, he says:

> We can investigate nature and develop a theoretical understanding of the world, but we cannot compare what we think we know with the truth to see how well we are doing. (1985: p69)

We have no access to the theoretical independent of the theory, thus how could we ever know if our theories are true? For we have no *view from nowhere* with which to enquire as to how successful we are. The matter can be put very simply: other than concerning the phenomena, how can we know that things are as the theory says they are? We can describe a theory, but we cannot describe the theoretical reality posited by the theory other than via that same theory. van Fraassen captures the problem:

> If two things are independently known and independently described, then it makes sense to claim that they are like or unlike in some respect, and the claim can be evaluated. But if only one is described, what is meant by a bare claim of similarity? How can that be non-trivial or informative -- how could it even be true or false? (1997: p511)

The realist claims that the theoretical reference of the theory is something beyond its correctness and adequacy in representing phenomena. In other words, for a given theory, which (we may suppose) does *represent* the phenomena correctly and adequately, there are still two possibilities: (a) that its theoretical terms do successfully refer, and it is thus *true*, and (b) that they do not refer and it is thus *false*. How could we ever tell which of these is the case?

§8 The No Miracles Argument for Scientific Realism

A different Hilary Putnam from the one who proposed the NMA described realism as 'an impossible attempt to view the world from nowhere' (1987: p28). Ernan McMullin, referring to theory-world correspondence, says that the realist

> would suppose that the structures of a theory give some insight into the structures of the world. But he could not, in general, say how good the insight is. He has no independent access to the world, as the antirealist constantly reminds him. (1984: p35)

Many realists deny that *unmediated* independent access is required, claiming that it is sufficient if there is mediated access via, for example, an inference to best explanation, and I will examine below the views of Bird as an example. However, as we have already discussed in §8.4, the antirealist will have no truck with IBE so responding to the antirealist's 'independent access' problem in this way will give little success. Nevertheless, this again highlights the centrality of IBE to the realist, for it is the only way that the independent access problem of correspondence truth can be circumvented – some method of confirmatory inference is needed. However, note that an inference to truth is never quite the same as unmediated direct access, a point I shall return to below.

Realists may object that they do not have to rely upon the actual *correspondence* theory of truth. For example, Chakravartty insists that

> the correspondence theory of truth is not a requirement of realism. Indeed, different realists hold different views regarding the nature of truth, and different views are compatible with realism. (2007: p201)

§8 The No Miracles Argument for Scientific Realism

But Chakravartty himself affirms that they cannot avoid the wider sense of *correspondence* in which, for the realist, theories are thought to correspond to the world. He notes that:

> it is important here to distinguish the general issue of correspondence between language and the world from the more specific idea of a correspondence theory of truth. (*ibid.*)

The semantic problem of how language refers to the world is seen in its purest form in the correspondence theory of truth. However, for the realist, theories *refer* to the world; are *about* the world; and their terms *refer* to items in the world. Consequently, realists who avoid the correspondence theory of truth still confront this more general problem of theory-world correspondence. Chakravartty points out that if theories are to yield knowledge of both theoretical and non-theoretical aspects of the world, as the realist believes, then issues of correspondence unavoidably arise, and thus the realist 'cannot avoid talking about correspondence in some form, even if it is by means other than the correspondence theory of truth.' (pp201-2).

The realist who accepts the correspondence theory of truth confronts the problem of independent access head on, but those who embrace other accounts of truth, or none, do not thereby evade that problem. In terms of the global realism/anti-realism debate, the issues surrounding the 'correspondence theory of truth' may well be of central importance, but in the more local issue of scientific realism/anti-realism the problem takes a different form, for the issue here is not about how our language corresponds to an external reality, but how a *theory* corresponds to a reality, parts of which we come to know about only via the theory – the independent access problem. All the questions concerning language that arise in the global case arise equally in our local case, but now we are confronted with this additional problem: the theory is said to correspond to some portion of reality, but, as stated earlier, absent IBE, the truth of this can be established only via that very theory. The quantum theory involves 'quarks' and if that theory literally corresponds to the world then the theoretical term 'quark' refers to, or corresponds to, quarks that are part of 'reality' in the same way that my chair is. However, absent IBE, this could be established only via the quantum theory itself, which would involve circularity. Why does this problem not arise in the global case? Because, I claim, although 'chair' refers to *that* chair, it does not do so via any theory. Indeed, unlike 'quark', 'chair' is not a constituent of a theory at all, but a term whose use is socially established to refer to an item that is multiply perceived by many agents. Furthermore, the *many* agents ensure that even if the sceptical philosopher suggests that 'chair' is part of a theory that I hold, I can still gain independent verification that there is a chair there from the inter-subjective confirmations of all the other agents. No such means of confirmation exists in the scientific case. Here the observable/unobservable distinction seems legitimate as the global case certainly concerns only those items like chairs that are readily observed and inter-subjectively checkable. References to unobservable phenomena arise only in the case of scientific theories.

Against this argument, we may take Bird as typical of those who deny that there needs to be any theory independent access to the quark in order to know that certain quark reports are true. In effect he demonstrates the way in which the realist notion of correspondence trades upon an implicit assumption of the legitimacy of IBE. Given a scientific statement S concerning world fact F, Bird denies that access to F has to be independent of S and says that

§8 The No Miracles Argument for Scientific Realism

> The truth of S may consist in its matching F. But it does not follow from this, that knowledge of the truth of S requires unmediated knowledge of that matching. Why cannot knowledge of the matching be mediated (for example, by *inference*)? We might think that the matching between S and F is the *best explanation* of the truth of propositions deduced from S. Or we might think that the fact that F is the *best explanation* of certain other facts. From which we *infer* that F exists, from which we in turn *infer* the truth of S. (2007: p82, my emphasis)

So confirmation of a correspondence between theory and world can be had by reliance upon inference to the best explanation – IBE, but as we saw in §8.4 the validity of such an inference is one of the central issues at stake in the debate between realist and anti-realist. Thus Bird's argument would be circular if deployed in support of the NMA, for if the realist is asked to justify the word 'true' in the NMA she will eventually have to resort to IBE in order to defend it.

The realist strategy is to get round the independent access problem for correspondence by resorting to an inference in the case of theoretical items, but is this really acceptable? In the case of observable items, as stated earlier, the correspondence is plain to see – it is established by simple ostension in an infallible way and thus the correspondence can support *truth*. However, as I showed in §8.4, inferences are fallible, particularly in the case of IBE. So surely a fallible correspondence can only support fallible possibility, not truth. Where Bird writes '... infer the truth of S' he is strictly only allowed '... infer the fallible possibility of S'. Consequently, resorting to IBE only supports correspondence *qua* fallible possibility, not correspondence *qua* truth.

§8 The No Miracles Argument for Scientific Realism

In his book on Kuhn, Bird employs another argument for correspondence truth that relies upon IBE:

> Why should knowing there is a match between theory and reality require having independent access to each? Consider this analogy: I know that there is a match between my key and the levers of the lock, not because I have independent access to the levers of the lock, but rather because the key opens the lock. (2000: p227)

This is again aimed at circumventing the independent access problem. However, it seems a poor argument, for it is clear that I do have independent access to the innards of the lock by virtue of the fact that I know it was made by some locksmith who used certain latching technology such that only my key will match this lock's particular latching pattern. More generally, I know that it is a lock and that a key is supposed to match only one lock. I may well use IBE but the inference is based upon my prior knowledge of locks – their purpose and roughly how they operate. This analogy could be made to match the scientific theory case only if I knew nothing whatever about the lock's internal construction, and didn't even know that it was a lock. In that case I clearly would *not* know there is 'a match' even if my key opened the lock. An alien on a hidden spacecraft above might have been observing, saw me putting the key into the lock, and independently caused the lock to open by some technology beyond human knowledge; or I might be just plain lucky; or all keys would open this lock, not just mine. In each case the inference that Bird recommends would lead to a wildly wrong conclusion.

Bird's argument seeks to draw an analogy between the lock and key which correspond to the non-theoretical terms of a theory, and the lock's hidden latching mechanism which corresponds to a theoretical term in the theory. However, the analogy fails because in the case of the lock Bird knows many background facts about the situation – what a lock is for, how locks are made to have unique keys which *match* them – indeed one might almost argue that the very terms *lock* and *key* are constitutive of the notion that the key uniquely matches that particular lock. It is the knowledge of these background facts that circumvents the problem of independent access. In the case of the scientific theory none of these background facts are available – the only facts known are those made available via the theory itself and the problem of independent access cannot so easily be avoided.

It seems that the claims of the NMA in terms of truth entail either reliance upon IBE, with, as we saw above, the circularity which that brings to the NMA, or an adherence to a problematic notion of correspondence.

To summarise, the NMA specifically, and realism more generally, both require the notion of correspondence. This can be exemplified in the correspondence theory of truth, though this is not essential. But that notion of correspondence suffers from the independent access problem and that can be circumvented only by utilising arguments involving IBE, which, I shall argue, render the NMA circular. Finally, in the light of these problems, and those associated with *approximate truth* some realists begin to look away from truth based definitions of realism altogether.[16]

---

[16] Bird (2007b) deplores the dependence on truth and verisimilitude in current realism, suggesting a conception of scientific progress not connected with truth, but with the accumulation of knowledge. He follows Williamson (2000) in holding not truth, but knowledge, to be the central, non-analysable, epistemic concept.

§8.6 Unconceived Alternatives

I have devoted an entire chapter (§7) to the underdetermination of theory by evidence and its relevance to the realism/anti-realism debate. However, here I think there is a specific issue concerning underdetermination that seems to undermine the NMA. I earlier suggested that the more grandiose claim of what one might call the 'classical underdetermination argument' – that there is an infinity of alternatives for any given theory – could be discounted as an example of radical scepticism. However, the 'unconceived alternatives' argument suggests that it is still the case that, for any of our current scientific theories, we have good reason to suspect that there may well be at least one currently unconceived alternative that would be incompatible concerning theoretical terms (in other words, such an alternative would give a different account of the world), but which will make precisely the same predictions concerning observable phenomena. Let us recall the second premise of the NMA, slightly rephrased:

NM2'   The only (or best) explanation for the success of a scientific theory is its truth or approximate truth.

Here success is characterised in terms of successful predictions – i.e. observed outcomes. However, if the *unconceived alternatives* version of underdetermination is accepted, then for our current successful theories there may be other theories that are incompatible concerning their theoretical terms, but which will make precisely the same predictions. Moreover, the induction proposed in §7 says that we have very good reason to think that there may be such alternatives, but because of the incompatibility of such alternatives they could not all be true. It might be argued that they could nevertheless all be approximately true, but here we again seem to be up against the vagueness of that term. For,

§8 The No Miracles Argument for Scientific Realism

by definition, these theories will be incompatible concerning their theoretical terms. If one said, for example, that continents float on plates, while another said they do not, could they both be approximately true? Or if one said there are quarks, but the other said there are not, could both be approximately true? The answer to these questions is indeterminate since no definition of 'approximate truth' exists that can answer such questions. I am therefore entitled to claim that if one or more theories all give the same predictions but are incompatible at the theoretical level, not all can be approximately true. Moreover, the central problem for the realist is to decide how to choose one such theory over another. Lipton says that for any set of observations

> there are many incompatible theories that would have had them. ... the truth explanation does not show why we should infer one theory rather than another with the same observed consequences. (2004: p195)

The only reply that can be made to this will be in terms of pragmatic properties of theories – simplicity, fecundity, generality, Occam, aesthetic considerations, etc. – all criteria of what Lipton calls *loveliness* (*ibid:* p59). However, none of these play any part in the NMA as stated.

As we saw in §8.4, IBE should not warrant inferring the best explanation at all costs – the best must be good enough to be better than no inference at all. Maybe the most likely explanation of the success of scientific theories is that continued success is just an inexplicable brute fact; or perhaps it is just as likely that a false theory should be successful as a true one;[17] or it could be that there are other competing explanations, as yet unconsidered.

---

[17] A possibility strongly suggested by the historical record.

§8 The No Miracles Argument for Scientific Realism

How *lovely* is the truth explanation? In drawing his distinction between loveliness and likeliness for explanations, Lipton conceives the loveliest explanation to be the one that would, if correct, 'be the most explanatory or provide the most understanding' (*ibid*: p59). *Prima facie*, the truth explanation thus has some loveliness, but underdetermination shows that for any set of observational successes, there may well be other incompatible theories that would have those same successes, and thus the truth explanation would apply equally well to all of these theories – for each theory, its being true would explain its success, and each of these explanations would seem equally lovely. For let us remember that we are not here considering the loveliness of the separate scientific theories, in terms of fecundity, simplicity, and so on, but the loveliness of the second level 'truth as explanation of success' theory as applied to each of these first level scientific theories. The truth of a *complex* theory is just as lovely an explanation of the success of that theory's predictions as the explanation that the truth of a *simple* theory provides. So the truth explanation gives us no criterion at all by which to infer one theory rather than another and thus is of dubious loveliness when this underdetermination problem is confronted.

To appreciate this point, it is important to distinguish the miracle argument, as second order explanatory inference, from the first-order explanatory inferences which scientists make. Those inferences, construed as inferences to the best explanation, do distinguish between theories with the same observed consequences, since not every such theory gives an equally *lovely* explanation of the evidence. This is one of the main strengths of Inference to the Best Explanation as an account of scientific inference. However, the truth explanation of NMA, applied to a particular theory, is distinct from the scientific explanations which that theory provides, for most of those scientific explanations are causal, but the truth explanation of NMA is not. The truth of a premise in a valid

§8 The No Miracles Argument for Scientific Realism

argument does not *cause* the conclusion to be true. The result is a very weak explanation that cannot itself show why one theory is more likely than another which has the same observed consequences.

The NMA captures the realist intuition that it would be miraculous if a successful theory were false. However, if we take these unconceived alternative theories seriously, it no longer seems miraculous that our successful theory might be false for the truth might lie among the underdetermined – unconceived – alternatives.

Let me illustrate this problem via quantum theory, perhaps the most predictively successful theory of all. Fine claims that quantum theory is itself the ultimate argument against realism:

> Realism is dead. ... Its death was hastened by the debates over the interpretation of quantum theory where Bohr's nonrealist philosophy was seen to win out over Einstein's passionate realism (1984: p83)

I will not argue for a position as strong as that, but I do suggest that quantum theory is a considerable problem for the NMA. Considered as a set of equations, quantum theory is enormously successful, and these equations underdetermine what is being said about reality as they leave open very different interpretations. In this case we can take the 'theory' to which NMA applies to be the sum total of the equations together with the interpretation of what is being said about reality, and then we have several different theories, all of which make the same highly successful predictions. However, these theories are radically different descriptions of reality, varying from Everett's *many worlds*, to Bohm's deterministic *hidden variable* theory. Which of them does NMA deduce to be true or approximately true? According to NMA we must

§8 The No Miracles Argument for Scientific Realism

take each of them to be true or approximately true since each of them gives predictive success. However, it is hard to see how all of them can be approximately true, for how could the Copenhagen, Everett, and Bohm theories *all* be approximately true? Again, the vagueness of 'approximate truth' confronts us, but it is surely the case that theories having radically different ontological commitments cannot all be approximately true. If the Everett theory is approximately true, then surely none of the others could be. Perhaps the only option for the realist here is to say that all of the interpretations are to be discarded, thus leaving a still successful set of equations, but that doesn't sound like a form of realism at all, but a variety of instrumentalism. A related way out of this tangle for the realist, but at great cost, lies in Worrall's *Structural Realism* (1989) which I discussed in §6.6.8. As I suggested there, this leads far away from the kind of scientific realism with which this essay is concerned.

One reply realists sometimes make here is to say that quantum theory is just a rather special exception which proves no general point against realism, which can still be regarded as applying to other theories. However, this won't do, for quantum theory exhibits extraordinary levels of predictive success, and if truth is not to be the explanation of that fact, then what is? The existence of just one theory that is not true or approximately true because of underdetermination, yet gives perhaps more predictive success, in terms of accuracy at least, than any other scientific theory, undercuts the entire credibility of the NMA and the truth explanation of success.

The problem of underdetermination seems a decisive refutation of NMA.

§8.7 Conclusions

The arguments presented in this chapter, together with the discussion of PI (§5 and §6) and the question of Unconceived Alternatives (§7) show that very significant problems face the NMA, and suggest that it can no longer be regarded as a serious support for realism for these reasons:

- There is a heavy reliance upon the term *approximate truth* which §4 has shown simply has not received the required clarification. This results in discussion of both the NMA and the PI argument being plagued with argument and counter-argument that centre around the uncertainty of this phrase.

- The PI argument shows that the historical record stands against the assumption of any link between the predictive success of a theory and either its approximate truth, or the successful reference of its theoretical terms.

- The NMA is itself a form of IBE and this again makes it ineffective and question-begging against anti-realists who reject this method of reasoning.

- Regardless of whether NMA adherents rely upon the correspondence view of truth, the NMA is reliant upon a general notion of theory-world *correspondence*. This is confronted by the classical problem that confronts all *correspondence* theories – the problem of independent access, and that can be overcome only by reference to IBE, a rule of inference which the anti-realist rejects.

- The problem of underdetermination means that for any empirically adequate theory, there may be other unconceived alternatives. Which of these would NMA decide is the bearer of approximate truth?

In the next chapter I will show that anti-realism can itself provide an excellent explanation for the success of science.

§9 Explaining the Success of Science

§9.1 Two Kinds of Question

The arguments presented against the No Miracles Argument (NMA) in the previous chapter show that whatever the explanation for the success of scientific theories may be, their approximate truth would not be such an explanation, let alone the best one.[1]

However, in order to give a fully adequate reply to the NMA it is not enough to have shown that realism is not in fact the best explanation of the predictive success of science. I must either deny that science is a success or offer some alternative account of that success, and preferably an account that anti-realism is able to offer, or at least be consonant with. For if no explanation can be offered then the 'miracle' jibe will ring true.

Let me recall the first premise of the NMA:

NM1 The enterprise of science is enormously more predictively successful than can be accounted for by chance.

The success of science is denied by social constructivists such as Latour & Woolgar (1979), and by sociologists of knowledge such as David Bloor and Barry Barnes.[2] However,

---

[1] In the previous chapter I established that it is *predictive* success that is referred to. Henceforward this can be assumed whenever it is not made explicit.

[2] See Bloor (1991), Barnes & Bloor (1982). According to the 'Strong Programme', the outcome of all scientific controversies, successful or not, should be explained by social factors. Whether these writers use the word 'success' in the purely 'predictive' sense is questionable, but I won't discuss that further.

as far as I am aware no anti-realist has denied it, and indeed, partly due to its positivist roots, scientific anti-realism has been distinguished by its acknowledgement of the predictive success of science. So perhaps the anti-realist should offer some explanation of this success. As we shall see, not only is the anti-realist able to offer such an explanation, but it is a better explanation than that which the realist offered, even if that had worked. For it not only explains the success of our current theories, but also explains the success of past discarded theories which are now regarded as false, a phenomenon about which the realist has nothing to say. As we saw in §6.4 in relation to the question of 'mature sciences', realism seems unable to give an account of science that embraces its past as well as its present.

However, I first need to address what seems like an ambiguity in this call for explanation. Let me recall (from p28n5) my definition of what it is for a theory to be successful:

> By describing a scientific theory as *successful* I mean that it enables us to make many more correct predictions than we would without it, which can also be taken as thus implying that it will enable successful interventions in the natural world.

Let us agree that current scientific theories are, broadly, successful. The demand for an explanation for why that is so could be taken in two quite different ways:

EX1 What features do successful scientific theories possess that could explain that success?

EX2 What is it about scientific methodology that explains the production of successful theories, or at least theories that are more successful than their predecessors?

§9 Explaining the Success of Science

Clearly the NMA assumes that we refer to question EX1 and that assumption is conducive to the realist who will argue that a successful theory must be connected with some *objective reality* in some way involving correspondence between the theory and that reality. She will then claim that it is that connection which gives the required explanation – in other words, the theory is true, or approximately true, and that is what leads to the success of the theory. Thus the formula for successful theories is that they must possess a common feature – an essence – they must be true or approximately true.

But question EX2 suggests that there may be something about scientific method and epistemology which itself leads to success – that successful theories result from something about the selection procedures which scientists use to sift their theories. A problem in this debate is that realists reply to question EX1 and complain that anti-realists generally reply to question EX2, thus missing the point and failing to reply to their demand for explanation. Leplin draws an analogy with someone watching the two Wimbledon finalists who asks why they are such great players. He says there are two ways to take this question – as a request to explain why Wimbledon finalists are great players, or as a request to explain why these particular individuals are such great players. The former request could be answered by mentioning the high competitive standards and prize money, but the latter by mentioning the training or special athletic gifts of these two particular players. Leplin continues:

> Analogously, to explain why the theories that we *select* are successful, it is appropriate to cite the stringency of our criteria for selection. But to explain why *particular theories*, those we happen to select, are successful, we must cite properties of *them* that have enabled them to satisfy our criteria. (1997: p9; original emphasis)

§9.2 Two Kinds of Explanation

I will refer to the realist reply to EX1 as the *truth* explanation – the truth (or approximate truth) of theories explains their success, though I might equally refer to it as the *essentialist explanation* – approximate truth is essential to success. My strategy will be to present the anti-realist alternative explanation for the success of science in terms of answering question EX2, and I will refer to this as the *selectionist* explanation for reasons that will become clear. I will then show that the anti-realist could not, in principle, give the kind of answer to EX1 that the realist demands, and that the realist's rejection of the selectionist explanation and insistence on a reply to EX1 is, in effect, question-begging against the anti-realist.[3]

Some anti-realists have attempted to answer the EX1 question, thus accepting the essentialist assumption but rejecting approximate truth as the essence. I shall examine Stanford's attempt at this later (§9.6), and will suggest that no realist would accept it as the kind of explanation they have in mind. Timothy Lyons (2003) also attempts to answer the EX1 question – he says it is not the truth of theories that explains their success, but their *empirical adequacy*, and realists do indeed reject this for not being the kind of explanation they seek. For example, Leplin (1987: p522) says that attributing the property of empirical adequacy to a theory merely 'restates its explanandum'. As Lyons himself comments, it isn't that the realist denies that a theory being empirically adequate is *an* explanation of its success, but rather that it isn't a good enough explanation because

---

[3] A claim substantiated in §9.5.

§9 Explaining the Success of Science

The realist will claim that the non-realist explanation simply pushes the question back further, and she will ask, What is the underlying reason why the theory makes only correct empirical predictions? (2003: p894)

Here Lyons notes that Musgrave (1988: p242) writes, 'One wonders how the empirical adequacy of a theory might be explained if not by postulating its truth'.

When confronted with the suggestion that the anti-realist could not, in principle, answer question EX1 the realist might reply – so much the worse for anti-realism. However, as we have already seen in §8.4, the anti-realist will say that the answers to EX1 offered by the realist are illusory since they derive, via IBE, from the putative link between success and truth, which is itself one of the major points of contention between realism and anti-realism. In addition, the realist assumes that all successful theories share an essential property which accounts for such success. The anti-realist not only rejects that assumption, but argues that it is responsible for the realist's inability to explain the success enjoyed by many false theories. Here it looks like a stand-off, the realist accusing the anti-realist of being unable to answer her question in the way she demands, and the anti-realist believing that the insistence on that kind of answer is question-begging, and that the realist's truth explanation would not have explained enough even if it had worked.

To answer question EX2 we will need to examine the evaluative procedures which scientists have used for identifying those theories which are likely to give successful predictions. The realist's NMA approach to explaining the success of science completely ignores this crucial aspect of science and thus offers no indication of how science comes by these theories that both realist and anti-realist acknowledge to be so successful. For realists to call their argument the 'No Miracles Argument' may seem ironic since it leaves the coming into existence of successful scientific theories an unexplained miracle.

§9.3 Scientific Methodology

§9.3.1 Introduction

A detailed analysis of the epistemology and methodology of theory testing would be a considerable undertaking, but perhaps here I can give a sketch of why science works so well at producing successful theories. The question can be stated in the following form:

> Why do many scientific theories enable us to make successful predictions, and to intervene in the world in pragmatically useful ways, so much more frequently and more accurately than, for example, the theories that were current two centuries ago?

Now we have a comparative question and in order to answer it we won't need to show that our theories are always, or even usually, reliable guides to how the world works. Rather, the question requires us only to explain why certain kinds of theories, produced by certain kinds of epistemic methodology, tend to promote predictive success more effectively than do other sorts of theories that use different approaches. In other words, why do some methods for the successful prediction of the world's behaviour lead to more reliable results than others? Laudan describes the need to explain the success of science as 'the challenge of explaining why certain modes of knowledge authentication produce more reliable results than others do' (1983: p97). I think that the right answer to this question is in terms of the methodology of science.

§9 Explaining the Success of Science

Of course, that methodology is a product of the way our world happens to behave. We live in a world that is very stable, relatively unchanging, and exhibits a high degree of regularity. Some philosophers (For example, Cartwright (1983) and Bhaskar (1978)) argue that these regularities can only be seen under laboratory conditions, and out there in the world they do not obtain. However, this view seems to insist that something can be a 'regularity' only if subject to laboratory levels of accuracy. The pressure on the surface of my body is pretty much the same every day; the temperature varies within relatively small limits and is always colder at night; when I put something down I can depend on it staying there unless moved; if I go to sleep for a while I can rely on things being pretty much the same when I wake up. If one thinks about it carefully, the extent of regularity in the world is extremely high. Things really do carry on behaving pretty much the same way every day. The laboratory isn't needed in order to expose or discover regularities, but only in order to enable their formulation with precision. This fact of stability and regularity, together with human intelligence, make it possible for methods of tracking and predicting these regularities to have evolved. Had our world been different in these respects the methods would either have been different or science as we know it would not have existed.

I will examine this under two distinct aspects. Firstly, the approach taken within science to the comparison of theories and the design of experiments for theory testing selects those theories that are more apt to make successful predictions – in other words, more apt to successfully 'capture the phenomena'. Secondly, the approach taken to theory succession[4] seeks to ensure that the later theory should capture as many as possible of the phenomena that the

---

[4] What van Fraassen calls 'Royal Succession' (2002: p115).

previous theory successfully captured, plus some more that it did not.[5] So we have two aspects of the methodology of science – on the one hand, the comparison of competitor theories and the design of experiments for testing those competitors, and on the other hand, the criteria for theory succession. Let me examine these two aspects of the argument in more detail.

§9.3.2 Theory Testing[6]

Let us examine two examples of situations where we need to compare different theories in terms of their predictive success and pragmatic usefulness. In the first example I find that one cold day my car won't start so it is taken to the repair garage. The next day is warm, and the mechanic tells me the car is repaired and now starts every time, because he has replaced the brakes.

When I complain that I didn't need new brakes, he replies that my car starts now, and that he has therefore cured the problem I complained of. He adds that he had three other cars with ignition problems, and his changing the brakes fixed them too. So he claims that replacing the brakes cured my starting problem, and as evidence he cites his claim that all the cars whose brakes were replaced now start perfectly. My protestations that the state of the brakes couldn't be related to the car's ignition are ignored, and he replies that this just happens to be my theory about how cars work, but he is a trained mechanic and has a different theory, according to which brake wear can cause poor ignition. In support he offers the fact that the car started well once the brakes were replaced. In reply I might point to a different explanation for

---

[5] Of course this is grossly simplified, and will be discussed in more detail in §9.3.3.

[6] This discussion owes much to Laudan (1983).

§9 Explaining the Success of Science

the sudden improvement in my car's ignition, namely the warmer weather. Because of this, if he expects to get paid, he needs to produce some empirical evidence which supports his explanation of the ignition's being improved by changing the brakes, rather than my weather based proposal. Moreover, if I can show that the ignition of other cars has improved dramatically with warming weather when no one changed their brakes, my case is won.

In effect, the mechanic and I are using different epistemic strategies for the evaluation of our beliefs. The mechanic clearly forms his beliefs according to a simple *post hoc ergo propter hoc* policy. However, I insist that more discriminating tests be designed in order to rule out some of the many theories his policy would support. In fairness, my theory is also supported on *post hoc* grounds, but I can point to improved starting performance in other cars, whose brakes were not replaced. Surely all who think about these two strategies for the evaluation of empirical claims must find it obvious which is more likely to lead to reliable beliefs – i.e. beliefs apt to lead to accurate predictions. The mechanic's failure to impose experimental controls on his causal claims will lead him to make less reliable predictions than I will. Some hypotheses which pass my kinds of tests may be mistaken, and some which pass his tests may lead to correct predictions. However, my strategy will produce conjectures which go wrong less frequently than his, and that is precisely what it is to say that one theory is more successful than another. One method of theory testing is more effective than another.

Now let's look at a more complex, and more typical, case. Consider the testing of a new drug believed to be good for curing back pain. To test it we might begin by asking a group of doctors to prescribe it for their patients who complain of back pain. Perhaps the results show that 60% of people taking the drug showed reduced back pain within a week. So now maybe we will think the drug a success. However, the test is badly designed since, for all we know, maybe 60% of patients who take nothing at all also report improvement within a week.

So in the next testing stage we need a more complex experiment. Maybe we might divide the patients into two groups, administering nothing to one group and the new drug to the other. Suppose it then emerges that 60% of those given the drug show improvement within a week, while only 20% of those given nothing show improvement. Well again we must be careful about our conclusions; for it may still be the case that the drug is of no therapeutic value at all despite the results. It may be that patients given any pill, even a sugar pill, will report some improvement due to the placebo effect. Because the control group was given nothing at all, the different results for the two cases might not be related to the drug being investigated.

So we re-design our test again. Now we give pills to both groups, but only the doctors know that the control group's pills are worthless. Suppose the results are that the group given the real drug shows a 60% improvement after a week while the placebo group shows a 30% improvement. Now the evidence for a real effect is increasing, but still caution is needed. Many other similar experiments have taught us that the doctors conducting such experiments might unconsciously transmit their knowledge and expectations to the patients in the experiment. Or the reverse could happen and the doctor's reports of symptom change may be influenced by their patients' comments. For a really strict test of the drug, we need to ensure that those doctors giving out the pills also don't know whether they are giving the real drug or the placebo. Such an experiment is devised. Of course this latter 'double-blind' method is now pretty well standard in drug and treatment testing.

§9 Explaining the Success of Science

The above description of successively improved experimental method shows transitions, first from an uncontrolled experiment to a controlled one, then from a controlled but non-blind experiment to a single-blind one, and finally from a single-blind to a double-blind experiment. For each transition good reasons can be given as to why the results of the later test should be more reliable than those of the earlier one. Those who study experimental method know the story well, but even those unfamiliar with scientific methods can surely see that, if our concern is to find out whether the drug being tested really works, procedures such as control groups and blind testing give us increased control over the many variables that enter into such situations.

This illustrates the way in which experimental method and theory choice contribute to the controls that are a part of 'scientific method'. However, there is another, equally important source of such control – the peer-review process. This is an example of a pragmatic, socially evolved component of scientific method. It has evolved over time as a sophisticated way of meeting the requirement for inter-subjective verification, and it demands a detailed statement of all experimental details and assumptions which can then be subject to rigorous scrutiny and/or attempted replication. Ladyman stresses the central importance of this process to an understanding of the success of science, saying that its uniqueness can be seen in its

> institutional norms: requirements for rigorous peer review before claims may be deposited in 'serious' registers of scientific belief, requirements governing representational rigour with respect to both theoretical claims and accounts of observations and experiments, … [we can] achieve significant epistemological feats by collaborating and by creating strong institutional filters on errors. (2007: p28)

§9 Explaining the Success of Science

Why do I associate this aspect of scientific methodology with the production of successful theories? Because the peer-review process assists in ensuring that successful predictions have indeed been made. For a peer-reviewer is not likely to make a successful objection to a scientific paper simply on the grounds that the reviewer happens to *believe* that paper to be untrue, but on the grounds that the paper entails phenomena that deviate from empirical findings in some respect.

The comparative reliability of both testing procedures and peer-review methods can be explained without needing any reference to claims about the truth-likeness of the theory in question. To see why science works, look not to scientific realism but to its rigorous peer-review process, together with a good textbook on experimental design, and there we will find no reference to the 'truth' of theories.

Moreover, such an explanation of the success of science can itself be tested. It predicts that individuals or societies which construct their beliefs without the kind of controls that we associate with science, will in general end up with beliefs that are less reliable than the beliefs of a 'scientific' culture. By 'less reliable' I mean that those beliefs will be less good at making accurate and useful predictions. This is not to say that non-scientific cultures can never have beliefs which are conducive to successes which have eluded Western science, since weak epistemic methodology may sometimes hit upon useful discoveries. Rather, the claim would be that the frequency of such discoveries should be lower in non-scientific cultures than in scientific cultures. This certainly seems to be the case, though I do not say that I have such evidence immediately to hand. However, my point here is that my proposed explanation for the success of science is empirically testable, unlike the realist explanation (see discussion on p247). The relevance of this is that many realists wish to claim that their realism is appropriate for a naturalistic philosophy of science – philosophy continuous with science, but here anti-realism is in a better position to offer such a naturalistic explanation, as compared to realism's explanation in terms of truth and correspondence – transcendent metaphysical notions that have nothing to do with the actual practice of science.

§9.3.3 Theory Succession

If we consider the reasons why a scientific theory is discarded and replaced, we find two. First, the earlier theory fails to account for some empirical observations – these may be wholly new observations, or they may have been known for a long time but ignored in the absence of any alternative theory. Now an alternative theory presents itself which does account for these observations that the previous theory was unable to accommodate. Second, a new theory is conceived which matches all the observational results of the previous one, so in that respect they are equivalent. However, the new theory makes some novel predictions which cannot be aligned with the previous theory. It is a requirement of the new theory that it should preserve the successes of the old one. So van Fraassen says:

> One of the main credentials for a new theory, offered as a rival to an old one, is that it be able to explain and preserve the successes of the old theory. (2002: p115)

Clearly van Fraassen thinks that a new theory wouldn't even be considered unless it explains its predecessor's successes. Kuhn would disagree, but I think that the disagreement does not affect the argument here. When a new theory replaces an older one the perfect case would be that the new theory captures *all* the phenomena captured by the previous theory, plus some additional ones. However, I concede to Kuhn the fact that immediately after a new theory has replaced an older one, it frequently cannot capture even all the phenomena that the previous theory could, let alone all the phenomena. The reason the new theory supersedes the old one is that it is believed to have the *potential* to capture more phenomena,[7] while the previous theory may have no more potential left. The new theory may be unable to capture some previously captured phenomena, but when the sum is considered of those phenomena it can capture, together with its potential for further phenomena, then it will be seen to have the potential to capture more phenomena than the previous theory.

Of course, there is also the phenomenon of 'Kuhn-loss',[8] making this not necessarily strictly the case. There may also be other pragmatic features of the successor theory that make it more attractive than the older theory, for example, elegance, simplicity, removal of significant *ad hoc* features, but these could be considered as simply contributing to the theory's potential for future success. Nevertheless, it remains the case that the successor theory must be considered to have the potential to be more successful in terms of the totality of what it is able to predict. This is a *sine qua non* of theory succession, and a theory that was not even potentially as successful as some previously accepted theory at capturing the phenomena would not be adopted. Consequently, if we look at successive theories, we can see definite progress in terms of their ability to make useful and successful predictions as to the observed phenomena, even if the new theory may require some years to achieve its full potential.

---

[7] See Kuhn (1970: p169).

[8] Kuhn (1996: pp107, 148, 169) points out that scientific revolutions can involve a revision in which not all the successful predictions of the preceding theory are preserved, and the later theory may even lack any explanation for a phenomenon that the earlier theory explained. An example would be Descartes' theory of vortices which explained why all planets orbit the Sun in the same direction, while Newton's theory does not.

§9 Explaining the Success of Science

In a sense, both of the reasons I gave for theory replacement amount to the same thing – there is a difference in the set of empirical predictions made by the two theories, and, allowing time for its potential to become actualised, the new theory successfully predicts more than the old. If this were not the case then the new theory would not be adopted.[9]

It is precisely this method of theory succession that ensures that our current theories are highly successful. For what does 'successful' mean here? It surely means that our current theory enables an extremely large number of successful predictions, predictions that enable us to make pragmatically useful interventions in the way things happen in the world.

So clearly, if a current scientific theory stands at the end of a long historical line of previous theories we should expect it to be successful. For it stands at the end of a long series of these theory successions, each of which made an improvement to the ability to make accurate predictions. One is inclined to ask – how could such current theories *not* be successful? An example of such a series of theory successions would be the road from Aristotle's cosmology to Einstein's, via Ptolemy, Copernicus, Newton. None of these cosmologies was completely without success, and some enjoyed enormous success over a very long period (the Ptolemaic system for over a thousand years). However, at each theory succession the potential number and extent of possible successful empirical predictions increased.

---

[9] Of course this greatly simplifies the full discussion of theory succession – see, for example, Kuhn (1996), van Fraassen (2002: ch.3).

§9.3.4 Conclusion: Experimental Method Plus Theory Succession

Two complementary themes emerge in this discussion. First, that scientific theory comparison is accompanied by experimental and peer-review methods that seek to rule out the intrusion of effects that militate against the power of the chosen theory to make good predictions. Thus, confronted with some new phenomenon, the experimental methodology which I have described will come into play and it is likely to achieve success in terms of successful predictions. Second, when a completely new theory is considered, usually because the existing theory is somehow deficient in its predictive abilities, the predictions made by the new theory must be at least as good as the old one, but also improve upon it. I submit that these phenomena that describe the practical methodology of science stand as adequate explanation for the success of current theories – they answer question EX2 with a *selectionist* explanation. I shall subsequently discuss the realist's reply to this and then go on to argue that EX2 is the only question the anti-realist *can*, or *should*, answer; that question EX1 begs the question against her; and that the selectionist explanation has considerably more explanatory power.

§9.4 An Evolutionary Explanation for the Success of Theories

Unsurprisingly, van Fraassen rejects the realist's NMA as the explanation of the success of theories. Indeed, he is sceptical of the need to offer any explanation at all.[10] Nevertheless, he

---

[10] He says: 'Science, apparently, is required to explain its own success. There is this regularity in the world, that scientific predictions are regularly fulfilled; and

§9 Explaining the Success of Science

offers his own explanation and evidently takes it to be so obvious that he keeps it extremely brief, just three short paragraphs in fact! He considers the question of why mice run from cats and says:

> But the Darwinist says: Do not ask why the *mouse* runs from its enemy. Species which did not cope with their natural enemies no longer exist. That is why there are only ones who do. ...
>
> In just the same way, I claim that the success of current scientific theories is no miracle. It is not even surprising to the scientific (Darwinist) mind. For any scientific theory is born into a life of fierce competition, a jungle red in tooth and claw. Only the successful theories survive – the ones which *in fact* latched on to actual regularities in nature. (1980: pp39-40)

Just as the current survival success of mice can be explained without the need for dubious speculations as to their hidden thoughts, similarly, the success of our current theories can be explained without equally dubious reference to their *truth*. I think that what van Fraassen proposes here is a brief and telling way of saying what I have said earlier. The combined methodologies of experimentation, rigorous peer-review, and theory succession, maximise the ability to make successful and accurate predictions, thus creating that 'jungle red in tooth and claw' in which only successful theories survive. Thus the success of current theories can be attributed to the fact that the unsuccessful theories have been eliminated

────────────── footnote continuation
this regularity, too, needs an explanation. Once *that* is supplied we may perhaps hope to have reached the *terminus de jure*?' (van Fraassen, 1980: p39).

§9 Explaining the Success of Science

(hence 'selectionist' explanation), and no further explanation is required.[11]

Lipton (2004, pp194-5) argues against van Fraassen's evolutionary/selectionist explanation. I think he is unsuccessful in his argument against van Fraassen for reasons I shall discuss below. However, notwithstanding that, his argument doesn't affect my discussion of this issue because his analysis assumes that van Fraassen offers his explanation as being *better* than the truth explanation, in other words, as a refutation of the NMA:

> The miracle argument cannot be defeated on the grounds that the selection explanation is a better explanation of observational success and so blocks the inference to truth as the best explanation. (*ibid*: p195)

Whether or not this was van Fraassen's strategy, it has not been mine as I have presented a detailed refutation of the NMA prior to this discussion of an alternative explanation. My argument isn't offered as refuting NMA and its truth explanation, as that has already been done, but as providing a genuine and powerful alternative explanation which fills the gap created by the refutation of the truth explanation. Below I will take up another aspect of Lipton's discussion of these issues that I wish to reject.

Not only is this selectionist explanation fully adequate to explaining why current theories are successful, but it has two

---

[11] In terms of the evolutionary explanation, Ladyman (2007: p73) suggests that the selectionist explanation is a *phenotypic* explanation, in that it gives a selection mechanism for how a particular phenotype (a successful theory) has become dominant amongst theories. The complaint is then that what is required is a genotypic explanation concerning the underlying properties of successful theories. The realist then claims that a theory being approximately true is such a genotypic explanation of its success.

§9 Explaining the Success of Science

other explanatory advantages which wouldn't be possessed by the truth explanation even if that was valid. First, it enables us to explain the discarding of once-successful theories, and here the mice analogy remains useful. Suppose the environment of mice changes such that the disposition to run from cats no longer explains their survival. Whatever is the new predator will dictate a new standard of success and thus the current way of explaining mice survival will need to be discarded and replaced. So it is with science. As a scientific domain develops, theories will be called upon to explain things they had not previously been expected to explain. Theories therefore constantly face new problems, resulting in once successful theories changing their status and being discarded. Here is an example: following Galileo's observation of Jupiter's moons there was a new pressure on theories to explain how planetary moons remain in their orbit. Prior to Galileo this was not an issue for the Ptolemaic theory, but new demands were now placed on theories, and this ultimately led to the Ptolemaic theory being discarded.[12] A theory once widely accepted as successful came to be regarded as unacceptable. The selectionist explanation seems better suited to explaining this ubiquitous scientific phenomenon. However, the truth explanation would have nothing to say about it since all the realist could say is that we were simply mistaken – the theory which we thought reflected the structure of the world does not after all, and that theory's prior success is left inexplicable.

---

[12] In pre-Copernican cosmology the Earth was the one unique centre of motion. According to Copernican theory, the Earth circled the Sun and the Moon circled the Earth, hence at least two centres of motion. And if Copernicus was right and the Earth was a planet, like Venus, Jupiter, etc, why did it alone have a moon? Galileo answered these questions: the Earth was not the only planet with a moon as Jupiter had four, so there was now another centre of motion – Jupiter. So Jupiter's moons supported the case for Copernicanism.

§9 Explaining the Success of Science

The second advantage of the selectionist explanation is that it can explain how multiple competing theories can each give predictive success. Take, for example, current quantum theory. This is often thought of as one theory with several different interpretations, but it can equally be viewed as several distinct theories all making the same empirical predictions (also discussed in §6.2.3 and §7.2.3). What could the truth explanation say here, for clearly they cannot all be true? Nor is it credible to describe theories as different as those of Everett and Bohm as both being approximately true descriptions of the world. On the other hand, several competing theories both being predictively successful presents no problem for the selectionist explanation of success because if they are all successful in making predictions, we would expect each theory to be accepted by some scientists which is precisely what we find.[13]

So it seems that the selectionist explanation is superior to the truth explanation, not only because the latter doesn't work as an explanation anyway, but also because the selectionist explanation fits the whole history of science. It does this by explaining how most theories go through periods of success but later come to be viewed as unsuccessful, and also how it is possible to have several theories all equally successful at the same time.

§9.5 The Realist Reply Refuted

Realists reject these proposed explanations for the success of scientific theories – answers to question EX2 – claiming that they do not offer any explanation of why *this specific theory*

---

[13] Wray (2007: pp87-88) gives an example of two theories giving very similar predictions, namely Copernicus' theory and the late renaissance version of the Ptolemaic theory. Many others could be cited.

§9 Explaining the Success of Science

is successful. Leplin says that if we ask why theories are successful

> we need an answer that goes beyond an explanation of why science in general produces successful theories; we need an answer that appeals to attributes that discriminate among theories. Why does *this* theory work, while others equally the products of diligence and preferred methods fail? (1997: p8)[14]

Lipton, similarly, complains in a manner reminiscent of Ladyman's genotypic/phenotypic distinction (see p298n11):

> If a club only admits members with red hair, that explains why all the members of the club have red hair, but it does not explain why Arthur, who is a member of the club, has red hair. That would perhaps require some genetic account. (2004: p194)

Many writers who discuss this issue find convincing this suggestion that the anti-realist has changed the subject and not answered the true demand for explanation – question EX1. I disagree, as I think that realists here beg the question against anti-realists, a claim which I aim to substantiate in this section. van Fraassen says the explanation for why mice run from cats is evolution. However, realists say they want an explanation for why *this* mouse runs from cats – i.e. some explanation of the underlying mechanism in *this* mouse that explains *its* running from cats. So they want an explanation in terms of facts about the brain of a mouse and its inherited dispositions, etc. Now consider transferring this kind of

---

[14] See also Psillos (1999: p96) for similar objections. The anti-realist Stanford (2000: p272) also makes this complaint.

§9 Explaining the Success of Science

demand to the case of theories – I claim that good experimental method, combined with a particular approach to theory choice over a long period of time must lead to successful theories – much the same as van Fraassen's claim. This is in line with the overall anti-realist claim that all there is about a theory for it to be successful is that it makes good predictions, and *nothing more*. If realists now demand an explanation for why *this* theory is successful, they are demanding to know what features *this* theory possesses that explain *its* predictive success. The only answer the anti-realist could give is to refer to the nature of the world we live in, with its high degree of stability and regularity, together with the historical process that this theory stands at the end of, and point to the steady improvement in experimental methodology and theory choice, and to the insistence on better predictive accuracy. However, that is merely to repeat the story I have told above and which the realist rejects. If more than this is wanted, the anti-realist can only shrug and wonder what kind of an answer is required. For the only further explanation she *could* give in terms of some mechanism specific to this theory would be to display the theory and show how it enables good predictions, perhaps by showing how the structures of the mathematics and of the observed phenomena are isomorphic. Since she will not believe there is anything else of a deeper metaphysical kind, what else could she say? However, clearly realists demand some other kind of reply to their question which is more consonant with their own assumptions, and the anti-realist has no such reply to give.

Let me put it another way: Suppose someone agreed with Mach – that theories are economical representations of the world's empirical facts, and nothing more; or, a closely related position, that theories are simply mathematical models that enable accurate representation of the world's empirical phenomena. Beyond what I have already said, how could *that* person reply to the demand for an explanation for why a

§9 Explaining the Success of Science

particular theory is successful? All they could say is that it represents the phenomena better than other theories do, that this set of equations does a better job than any other. From that person's viewpoint such a reply would be sufficient and no more could be said; asking for more would be equivalent to demanding that they completely change their viewpoint. This would be question-begging because the realist would be assuming there is common ground as to the very possibility of the kind of explanation she demands.[15] The following paragraph will make this clearer.

What may motivate the realist demand for further explanation here is this: as I have conceded, an important component in the anti-realist explanation of the success of science is just the way the world happens to be. Central to the anti-realist explanation is the fact that the world is a stable and regular place. Were that not the case there would probably be no successful science of the kind that we have, or possibly no science at all. Conversely, the world's stability and regularity, combined with the intelligent curiosity of *homo sapiens* ensures a successful science. However, the realist wants to know why that regularity and stability is the case – why the world is regular in the way that it is. This is highlighted in Lipton's discussion of van Fraassen's EX2-type selectionist explanation which I discussed above. For Lipton (2004: p194) argues that the truth explanation explains why 'theories that were selected on empirical grounds then went on to more predictive successes', whereas the selectionist explanation does not. Here is the core of his complaint:

---

[15] A.W. Sparkes says: 'An arguer begs the question when he treats the matter under dispute as if it were common ground.' (1991: p100). Douglas Walton's discussion of question-begging also supports this view (1991: p13).

> Constructive empiricism assumes that scientific canons of induction yield theories that will continue to be empirically successful in new applications, but it does not explain why this should happen. (*ibid.*)

Lipton deems the truth explanation superior because it supplies this additional explanation, and here is the clear presentation of what I believe lies at the heart of realist dissatisfaction with any EX2-type explanation – the belief that it fails to respond to this other problem that always concerns the realist in any engagement with empiricist anti-realist proposals. However, the latter rejects the legitimacy of such a question, believing it represents a demand for explanation that goes beyond what is legitimate and can lead only to metaphysical speculation. Here we reach brute fact and what van Fraassen calls the 'Limits of the Demand for Explanation' (1980: §2.4). We reach the point where acceptance of further demands for explanation may be rejected according to the two empiricist principles I adopted from van Fraassen (See p23n22):

ii) A rejection of demands for explanation at certain points.

iii) A rejection of explanation by postulation.

Many realists express Lipton's complaint by saying that the anti-realist vision of science leaves it dependent upon a series of flukes, but Blackburn gives a good empiricist reply to Lipton concerning the question of why we should assume a theory will continue to be successful:

§9 Explaining the Success of Science

> If it is surprising, or lucky, that the patterns of events are simple enough for us to catch onto them and predict and control other events by their discovery, then it is also surprising or lucky that whatever reality is responsible for them is itself such as to issue in this simplicity. (2002: p115)

In other words, the realist's putative explanation of the regularities merely pushes the question further back, leaving it, in effect, unanswered. The empiricist's rejection of such explanation by posit goes to the very heart of her philosophy. Now we see substantiated the accusation of question-begging, as it is clear that the realist insistence on a reply to EX1 presupposes the legitimacy of the demand for explanation of the world's regularities, but the attitude taken toward such explanations is precisely what divides realists and empiricist anti-realists. My case is that the realist demand for a reply to the EX1 question is question-begging because it presupposes a common attitude on this fundamental issue.[16]

The anti-realist's selectionist explanation is of course accompanied by the observation that the world is actually highly regular and that as long as that continues to be the case then science will continue to be successful. In his discussion of this issue van Fraassen says:

---

[16] I believe that the selectionist explanation is successful for anti-realists of all stripes. However, my accusation of question-begging is substantiated only for the empiricist variety. I leave non-empiricist anti-realists to fend for themselves. The discussion of Stanford in §9.6 shows they may encounter difficulties.

§9 Explaining the Success of Science

> what explains the fact that all observable planetary phenomena fit Copernicus's theory (if they do)? From the medieval debates, we recall the nominalist response that the basic regularities are merely brute regularities, and have no explanation. So here the anti-realist must similarly say: that the observable phenomena exhibit these regularities, because of which they fit the theory, is merely a brute fact, and may or may not have an explanation in terms of unobservable facts 'behind the phenomena' – it really does not matter to the goodness of the theory, nor to our understanding of the world. (1980: p24)

The question of why *this* theory gives success can be answered only by adverting to properties of *this* theory that explain its success in more detail, and for the anti-realist that can only amount to a description of how the overall content of the theory leads to successful results. An analogy may be in order here. Some anti-realists will conceive of theories as models that are, in some respect, isomorphic with the observed phenomena. Suppose a model railway enthusiast buys a new electric train which he values as being a very good model of the real train. If someone demands to know why it is considered a good model, what can our enthusiast say in reply? He can only point to the model train, and perhaps display a photograph of the real train, and insist that it simply *is* a good model, and that the people who made it designed it precisely so that it *was* a good model.

I think that some writers fail to understand the force of van Fraassen's (and my) point here. For example, even the anti-realist Stanford says:

§9 Explaining the Success of Science

> Explaining the [predictive] success of a theory by appeal to its empirical adequacy [as opposed to its truth] is, in essence, to explain why some of the observational consequences of a theory are true by pointing out that all of its observational consequences are true; (2000: p268)

Stanford enlists Musgrave's support with this quote: 'This is like explaining why some crows are black by saying that they all are' (Musgrave, 1988: p243). Stanford finally concludes:

> van Fraassen gives us no reason for ending our search for explanations with empirical adequacy, and no justification for refusing to answer the question at just this point. (2000: p268)

However, it isn't that van Fraassen *refuses* to answer the question, but rather, as I have tried to show, that he cannot give the *kind* of answer that is demanded by the realist without, in effect, giving up a central plank of his empiricism – the refusal to give explanations by postulate. Indeed, *nobody* could answer the question unless they embrace the realist assumption.[17] For what can 'explaining the success of a theory' mean if one rejects the postulation of a 'hidden reality' standing behind the observations, as does, for example, van Fraassen? The anti-realist may maintain that there is nothing 'behind' the phenomena and the regularities we observe are all simply 'brute', or she might say, with van Fraassen, that we should be agnostic as to such a hidden reality. In either case, what kind of explanation could Stanford be demanding of such a person? What further explanation could be given of that fact other than to simply *display* the

---

[17] Stanford is not an empiricist, therefore the realist demand for an answer to EX1 is legitimate against him.

theory? Of course, if one acknowledges such a 'hidden reality', as the realist does, then a whole range of 'explanations' become available in terms of truth, convergence, etc. The anti-realist story can certainly be criticised, but not by posing a question and then insisting that the only acceptable reply is of a kind that itself assumes realism. There is no point in demanding an explanation of a kind that could be given only by a realist, and which can only be intelligible within the realist's framework of assumptions. When the realist demands an EX1-type answer to the question of why science is successful she is implicitly presupposing the realist framework, and is thus question-begging.

§9.6 Stanford's Proposal

Nevertheless, Stanford, himself an anti-realist (though no empiricist) believes he can give an EX1-type explanation, and in attempting to do so he provides an interesting twist to the argument. He has argued that anti-realists like van Fraassen have failed to give an adequate explanation for the success of science, and I have argued that nobody except a realist *could* give the kind of explanation that Stanford says is required. He reasons thus: consider a request for an explanation of the predictive success of the Ptolemaic system of epicycles. This would be answered by pointing out how closely its predictions come to those of the Copernican hypothesis that superseded it. Stanford invites us to call the relation between the two theories, *predictive similarity* of the Ptolemaic system to the Copernican. He then suggests that an appeal to predictive similarity is the right place to end demands for explanation of the predictive success of the Ptolemaic system. For if now asked why does this relation of predictive similarity exist between the two systems, the only answer would be to direct the questioner to the details of the Ptolemaic system itself, to see how its specific predictions arise from within its theory. If this doesn't satisfy the questioner then the only answer left is 'a puzzled look and a shrug' (2000: p273) – no further explanation of what intrinsic feature of the theory enables it to be successful is appropriate or possible.

Stanford now proposes that this relation of predictive similarity can be generalised and can be claimed to exist between every predictively successful theory and whatever may be the ultimately correct theory, resulting in this proposal:

> the success of a given false theory in a particular domain is explained by the fact that its predictions are (sufficiently) close to those made by the *true theoretical account* of the relevant domain. (2000: p275, my emphasis)

In other words, when confronted with a demand for an explanation as to why a theory is successful, he agrees to the realist's insistence on construing that question as being of type EX1, and not EX2 as van Fraassen advocates. That calls for an answer in terms of facts about that actual theory, not about the mechanisms that gave rise to that type of theory; and his answer is then – because the predictions made by the theory are very close to those that would be made by what he calls 'the true theoretical account'. He doesn't make clear what 'true' means here, and an initial reading makes it look like the realist's 'true' with the theory in question being approximately true. However, Stanford himself holds an instrumentalist position, so this is unlikely to be what he has in mind. The charitable reading of Stanford here is that he thinks that the 'true theoretical account' would be some ultimate theory which succeeds in giving absolutely perfect predictive ability, and thus *truly* describes the behaviour of the world (captures the phenomena).

§9 Explaining the Success of Science

This certainly has the surface appearance of being an anti-realist reply to the EX1-type question, but I am doubtful that the realist would be any happier with this reply than with my refusal to answer EX1 at all. For Stanford's reply and my refusal to reply both refer only to the phenomena, and make no reference to approximate truth or successful reference, and it is clear that no reply which fails to refer to these will satisfy the realist.

In addition, I think this 'surface appearance' is deceptive and that, in reality, Stanford fails to answer the EX1-type question. Let me grant that there is some theory TTA that is the 'true theoretical account', and that we have a successful theory T which has this relation to TTA of 'predictive similarity'. Now Stanford claims that, when asked why T is successful, his reply – that it has the relation of 'predictive similarity' to TTA – is a reply in terms of the theory T itself, and thus is of type EX1, as the realist demands. However, his reply reveals nothing intrinsic to theory T other than the fact that it makes similar predictions to theory TTA, but we already knew that. Theory T is successful; so the realist explanation is that T is approximately true. This is the kind of explanation the realist wants when he demands an EX1-type reply, but Stanford's explanation is simply that T gives the same predictions as would the 'true theoretical account', whatever that may be. However, to say that is to say nothing since it is just a tautology because, by definition, T gives successful predictions, so does TTA, so how can merely quoting that fact stand as an explanation for the success of T? I cannot see how a reply like this could be the kind of EX1-type reply the realist seeks. I thus conclude that Stanford fails to offer anything that contradicts my claim that when the realist demands the EX1-type explanation she presupposes the realist conceptual framework. When I claim that the anti-realist cannot give an EX1-type explanation, this is because, in principle, no such explanation is available outside of the realist framework of assumptions.

§9.7 Conclusions

Having refuted the NMA, I have now shown that anti-realism can explain the success of science in terms of the methodology of experiment, peer-review, and theory succession, viewed from an evolutionary standpoint. Realists object that this selectionist explanation doesn't explain what they want explained, but I have argued at some length that they are thus making a question-begging demand for the kind of explanation that only has any validity within the framework of realist assumptions. The realist's demand for an EX1-type explanation presupposes that realist framework and assumes that successful theories must all have something in common – they must be true or approximately true. If the anti-realist answered the EX1 question in a manner acceptable to realists she would no longer be an anti-realist! The anti-realist has a fine explanation for the success of science relating to such practical matters as design of experiments, theory choice, and theory succession. Moreover, this also explains how the entire historical scientific enterprise has yielded theories which are successful and whose total success increases over time, a fact for which the realist offers no explanation.

Realism seeks to explain scientific success in terms of a transcendent *truth*, but there is a very good explanation which is rooted in the nature of science as a social process that evolves over time, always oriented toward the making of successful predictions which facilitate pragmatically useful interventions in the world. This is the selectionist explanation of the success of science. It also succeeds in giving us a vision of scientific progress as the steady accumulation of greater ability to make those successful predictions. Realism seeks to portray scientific progress in terms of convergence upon truth, and that picture has been refuted. The selectionist approach to explaining the success of science also gives us a new and more empiricist/pragmatist conception of scientific progress.

§10 Conclusions

It is time to draw together the threads and summarise what I take this essay to have achieved. The fundamental issue which divides scientific realists and anti-realists is the question of the existence of the items referred to by the theoretical terms of scientific theories, the various attitudes taken by realists being affirmative, and anti-realists being agnostic or sceptical. Realists frequently state this in a slightly disguised form: they claim that theories are true or that science aims for true theories.[1] However, since empirical truth is a *sine qua non* of theories, those claims amount to claims concerning the truth of the theoretical commitments of theories, which in turn amount to existence claims for the items referred to by those theories.

What I have attempted to show in this essay is that the arguments given by realists in favour of those affirmative attitudes are unconvincing, and so are their arguments *against* the agnostic or sceptical attitudes of anti-realists. In other words, realists have failed to make the case for their position, and have also failed to refute the anti-realist's most important arguments.

I examined various formulations of scientific realism that can be found in the literature with a view to distilling the essential claims of a metaphysical, epistemic, semantic, and axiological kind. Most formulations of scientific realism can be expressed as some combination of these four claims, together with a claim for the correspondence theory of truth:

---

[1] Throughout this concluding chapter 'truth' can be taken to include the phrase 'or approximate truth'.

RC1 Most of the entities referred to by the theoretical terms of well-established current scientific theories exist mind-independently and have most of the properties attributed to them by science.

RC2 Scientific theories are typically approximately true and more recent theories are closer to the truth than older theories in the same domain.

RC3 Theoretical terms in scientific theories should be thought of as putatively referring expressions.

RC4 The aim of a scientific inquiry is to discover the truth about the matter inquired into.

RC5 The correspondence truth claim – that the correct way of understanding truth is via the correspondence theory.

I specifically set aside RC5 as beyond the scope of this essay, and, partly to simplify matters, I accepted RC3.

As we have seen, most scientific realists fall into two distinct groups. Firstly, there are those whose affirmative attitude amounts to the belief that our current scientific theories are, on the whole, true. They believe that most of the theoretical terms of scientific theories do succeed in referring to actually existent items and that the descriptions of those items given by those theories are largely correct, the words 'most' and 'largely' accounting for the move to approximate truth instead of truth. They also believe that successive scientific theories in any given domain converge upon truth and successful reference. I named this position Convergent Ontological Scientific Realism (COSR), comprising the conjunction of the two claims RC1, RC2.

The second group take quite a different approach. They claim that COSR is not the right way to formulate scientific realism because, they say, we should not assume that most of our

§10 Conclusions

current scientific theories are true, and some degree of scepticism is warranted. Many of those theories may be incorrect, but they claim that even if that is so, they can still be realists. For they say that the correct expression of the positive attitude I have referred to lies in claiming that science *aims* for the truth. Now for someone who finds it useful to talk of the 'aims of science', to say that science aims for the truth regarding the observable would hardly be a significant claim, for that would just amount to 'capturing the phenomena', and all sides of the debate accept that as a priority within the sciences. So clearly these realists are, in effect, claiming that science aims for the truth regarding the theoretical terms of its theories. I named this position axiological scientific realism (ASR).

What is common to both of these varieties of realism is the central place they give to truth and they both revolve around that notion together with its associated approximate truth. COSR makes it the primary product of science, and ASR makes it the central motivating concern. Moreover, I do not think that I have unfairly selected an unrepresentative sample of the total scientific realist population. On the contrary, I think it is fair to say that the two positions I have dealt with represent the overwhelming majority of realists, to such an extent that it would be reasonable to say that, with few exceptions, current scientific realism just *is* one or other of these two positions.

As I have shown, the truth which is so central to these realist claims is truth concerning the theoretical commitments of theories. The central problem faced by both of these varieties of realism is their inability to demonstrate, in a manner that might plausibly be persuasive to anti-realists, a connection between that theoretical truth and the acknowledged empirical success of science. That theme has recurred throughout this essay.

§10 Conclusions

In §3, p54 I attempted to undermine ASR, first by stating the pluralist position, that reference to *the sciences* would be more meaningful than to *science* simpliciter. I then argued that axiological claims in general are not illuminating and tell us less about the sciences than about the beliefs of the philosophers who propose them. I then went on to present several difficulties for the axiological claims of scientific realists in particular, and one of these concerned the problem of the absence of a connection between scientific method and theoretical truth. Scientific method is of course closely related to empirical success. In addition, the fact that scientific method seems particularly oriented to the attainment of empirical success makes it hard to believe that the central concern of the sciences could nevertheless be theoretical truth. The main conclusion of that chapter was that, despite the axiological scientific realist's disdain for COSR, her axiological claims strongly push her towards that position. Consequently, the remainder of the essay has been entirely concerned with COSR.

In §4, p94 I argued that the ubiquity of the phrase 'approximate truth' in all the claims of scientific realism made it virtually impossible to test those claims because, despite fifty years of research, the notion of approximate truth is still not sufficiently defined to enable realist claims to be determinate and testable. I also examined the more recent attempts by Psillos and Chakravartty to establish an intuitive notion of approximate truth and showed that such a notion is unable to do the work which realists need it to do.

In §5, p116 I presented what is generally known as the Pessimistic Induction argument (PI) in a deductive form aimed at demonstrating that realists have no basis on which to assume *any* link between empirical success and theoretical truth. In §6, p138 I defended the PI against an extensive range of arguments that realists have brought against it. At the end

§10 Conclusions

of that chapter I could reasonably claim that the PI has not been refuted and that realists have therefore failed to establish the link between empirical success and theoretical truth which they need.

In §7, p207 I examined underdetermination claims as they have conventionally been presented, and suggested that they were ineffective as arguments against realism, but that the problem of unconceived alternatives (PUA) does present a serious problem for realism. I also defended the PUA against several realist counter arguments. The conclusion was that it undermines our grounds for believing that the theoretical proposals of our current theories are true, and thus both RC1 and RC2 are undermined.

In §8, p244 I examined the realist's favourite argument – the NMA – and showed it to be question-begging in its use of a form of IBE which no anti-realist would accept. I also gave several additional arguments against it:

- its reliance upon the indeterminate term 'approximate truth'

- the fact that the PI argument refutes the NMA assumption of a link between theory predictive success and approximate truth

- the reliance of the NMA upon a general notion of theory-world *correspondence* which is itself reliant upon IBE, which the anti-realist rejects.

In §9, p281 I gave an alternative 'selectionist' explanation for the success of science. This is important since the rebuttal of the NMA is complete only if an alternative non-realist

§10 Conclusions

explanation of the success of science is offered. I attempted to show that the realist's rejection of this kind of explanation is question-begging because it assumes there is common ground as to the legitimacy of the kind of explanation demanded by the realist. Any anti-realist who succeeded in providing the kind of explanation which the realist demands would no longer be an anti-realist. I also showed that central to realist's dissatisfaction with this non-realist explanation is their rejection of the empiricist's acceptance of the world's regularities as brute fact. The realist demands an explanation. This deep division between realist and anti-realist accounts for their difference as to the acceptability of the non-realist explanation of the success of science.

There are three main conclusions. Firstly, that the realist fails to mount a convincing case for her position. Secondly, the realist fails to offer adequate replies to the various arguments offered by anti-realists. Moreover, these conclusions apply equally to both kinds of realism which I have examined. The third conclusion follows from these two – that scientific anti-realism is a perfectly reasonable and rational position to hold.

I could go further and say something like 'Such conclusions licence the view that scientific realism, as currently formulated, is a mistake, and the work done in this essay is a suitable prelude to making out a case for anti-realism'. I do believe both of these propositions. However, I want to suggest that it may be wrong to think the scientific realism/anti-realism debate ever could have reached a determinate "I'm right, you're wrong" kind of conclusion. Perhaps both realism and anti-realism are best seen as 'stances' in the sense discussed by van Fraassen (2002). I suggest that both scientific realism and anti-realism are not the kind of positions that can be refuted, just as van Fraassen showed that materialism isn't either (pp49-60). All three are positions that will always survive and regroup in the face of severe

§10 Conclusions

problems because they represent basic attitudes people take towards the world, and the attitudes always survive even if their precise formulations have to change.

Many times in this work we have seen central tenets of each side simply completely rejected by the other. Two examples: the anti-realist rejects IBE and no argument based upon it will have any persuasive force. Conversely, the anti-realist insists on not continuing the demand for explanation beyond what can be verified in experience, and the realist rejects this principle as being unreasonable. Can the acceptance or rejection of these two major principles – IBE and 'no explanation by posit' – be argued about? Perhaps so, but probably not with profit as it is likely that any such argument will terminate in the confrontation of two great philosophical stances – what one might call the empiricist stance, and its opposite, whatever one decides to call that.[2]

Another major theme that has run right through this work has been the question of a link between empirical predictive success and theoretical truth. In attacking realism, I naturally believe that there is no such link, but what is at the bottom of the great gulf between realist and anti-realist over this is again not so much beliefs, but different attitudes. For the anti-realist what matters most is the historical record. She sees science as a social practice, an activity with a glorious history which speaks against the existence of such a link. However, the realist elevates current science as the paradigm example of how to acquire knowledge (i.e. truths) and is not interested in the history of science except insofar as it could be seen as pointing towards the present. For realism, truth just *is* the regulative ideal of science and insofar as previous science may have been wrong by the standards of current science, well so much the worse for it.

---

[2] For example, 'realist stance' or 'metaphysics stance'.

§10 Conclusions

Considered thus, how could a reasoned conclusion be reached between two such opposed attitudes? On the one hand current science is put on a pedestal and eulogised, and seen as either achieving, or seeking, a transcendent truth – as discovering knowledge of a hidden realm. On the other hand, science is viewed as a historical social process rooted in the varying pragmatic needs of people. Given two such almost incommensurable views we should perhaps be grateful that discussion is possible at all!

Glossary of Terms and Abbreviations

Abstraction
An abstract description of a system leaves out a lot but is not intended to say things that are literally false. An idealised description of a system is a description that fictionalises in the service of simplification. The idealised description is often presented verbally as a description of a real system, but isn't a description that is literally true. Both idealisation and abstraction are an important component in scientific methodology, particularly regarding the construction of models, and this is reflected by the extensive philosophical literature.

Approximate Truth
In general, where I use the phrase approximate truth, the phrases 'very close to true', truth-like, or verisimilar may be taken as equivalent. These are discussed in more detail in §4.

Axiological Scientific Realism
These are claims of the form 'science aims at Φ' and are discussed in §3.

Central theoretical terms
This phrase refers to the theoretical terms (q.v.) of a theory that are of fundamental and central importance to that theory.

Confirmation holism
The view that confirming evidence flows to the whole of a theory, and not just to a part.

Constructive Empiricism
The name used by van Fraassen to describe his anti-realist position which claims that 'Science aims to give us theories which are empirically adequate; and acceptance of a theory involves as belief only that it is empirically adequate.' (1980: p12)

Convergent Ontological Scientific Realism
The conjunction of claims RC1 and RC2 (see §2.4)

Empirically adequate
A term of art used by van Fraassen: "A theory is empirically adequate exactly if what it says about the observable things and events in this world, is true – exactly if it 'saves the phenomena'" (1980: p12)

Global Realism/Anti-realism
Sometimes referred to outside this essay as *metaphysical* realism/anti-realism, a wider debate typified by the work of Dummett.

Idealisation
See Abstraction above.

Instrumentalism
Broadly, the position that scientific theories should not be taken as truth evaluable, but as means of codifying what happens, and hence (allowing for induction) of predicting what will happen. The position holds that theories do not correspond to the world but are pragmatically useful devices that enable the world to be predicted. It follows that instrumentalists deny the literal existence of all theoretical items, hence that van Fraassen is not an instrumentalist.

Internal Realism
A position proposed by Putnam (1981, 1987) when he forsook his earlier 'metaphysical realism'. On this view, we can make sense of notions like truth only by employing concepts like justification that are 'internal' to our non-philosophical practice. Thus the truth of scientific propositions would be relative to their means of justification, which would be the theory from which they derive. This is often interpreted as a version of anti-realism.

New Induction
This is the phrase used by Kyle Stanford to refer to his unconceived alternatives induction argument.

Glossary of Terms and Abbreviations

Observable/unobservable
It is difficult to give a simple definition that avoids contention, but see §2.2.

PUA induction
I use this phrase to refer to my variant of Stanford's New induction.

Referring term
A theoretical term (see below) can sometimes be taken as *referring* to a putative existent. This need not always be the case – for example, an 'inertial frame' is a theoretical term but nobody thinks it refers to anything that literally exists. On the other hand, the theoretical term 'phlogiston' was intended to refer.

Referring theory
If the central terms of a theory all successfully refer then we may say the theory is a referring theory, or that the theory refers or the theory successfully refers.

Selective truth realism
A term used by me to describe the realist gambit that assumes that evidence bestows confirmation upon only parts of a theory. See §6.6

Successfully refers
If a theoretical term does refer to an actual existent item, then I shall say that the theoretical term successfully refers or sometimes just refers. Clearly, in a trivial sense, all such terms do successfully refer, even if it is only to a concept or a mathematical object of some kind.

Successful theory
By describing a scientific theory as successful I mean that it enables us to make many more correct predictions than we would without it, which can also be taken as thus implying that it will enable successful interventions in the world. I also say that such a theory has predictive success.

Tacking Problem
Also known as the problem of irrelevant conjunction, according to which, if evidence E confirms theory T then it

will also confirm theory T + C where C is any further claim that doesn't undermine E's confirmation of T. Thus if confirmation holism was unconditionally true then E *would* give spurious confirmation of C.

Theoretical item

It is misleading to say that the theoretical terms of scientific theories necessarily refer to entities, as they can also refer to properties, events, or processes. An example would be the Big Bang theory, whose theoretical referent is best described as a process or event. I use the word item, or sometimes theoretical item, to refer to the putative referent of a theoretical term, and 'item' may be either entity, property, or process.

Theoretical term

See discussion in §2.2.4.

I also use the following abbreviations:

ASR: Axiological Scientific Realism (See §2.3.4 and §3)
CE: Constructive Empiricism (see above)
COSR: Convergent Ontological Scientific Realism
IBE: Inference to the Best Explanation
NI: New Induction (see above)
NMA: No Miracles Argument
NOA: Natural Ontological Attitude
OSR: Ontic Structural Realism.
PI: Pessimistic Induction. Sometimes referred to as the Pessimistic Meta-Induction, 'meta' referring to the fact that it is about science and its inductive methods, rather than within science itself.
PUA: Problem of Unconceived Alternatives (see above).
UTD: Underdetermination of Theory by Data (see §7).

Appendix 1: Varieties of Scientific Realism

For each of the following philosophers my aim is to draw out their metaphysical, semantic, epistemological, and axiological claims, and also whether or not they advocate the correspondence account of truth. Note that I do not examine the overall views of each person named, but merely highlight specific formulations of scientific realism which they have proposed. For want of a better ordering I have placed them in alphabetical order of surname.

A1.1 Richard Boyd

Boyd says this:

> (i) Theoretical terms in scientific theories ... should be thought of as putatively referring expressions; scientific theories should be interpreted realistically.
> (ii) Scientific theories, interpreted realistically, are confirmable and in fact often confirmed as approximately true by ordinary scientific evidence interpreted in accordance with ordinary methodological standards.
> (iii) The historical progress of mature sciences is largely a matter of successively more accurate approximations to the truth about both observable and unobservable phenomena. Later theories typically build upon the (observational and theoretical) knowledge embodied in previous theories.
> (iv) The reality which scientific theories describe is largely independent of our thoughts or theoretical commitments. (1983: p1)

(i) is a semantic claim concerning reference.

(ii) is an epistemic claim concerning approximate truth.

(iii) is an epistemic claim concerning convergence.

(iv) is a metaphysical claim concerning mind independence.

A1.2 Anjan Chakravartty

Chakravartty starts out with a definition of scientific realism in semantic and metaphysical terms:

> Scientific realism, to a rough, first approximation, is the view that scientific theories correctly describe the nature of a mind-independent world. (2007: p4)

Elsewhere, however, he characterises realism as:

> the view that our best theories yield approximately true descriptions of both observable and unobservable aspects of the world. (2008: p150)

This seems to involve the ontological, semantic, and epistemological aspects, and he is more explicit about this:

> Ontologically, scientific realism is committed to the existence of a mind-independent world or reality. A realist semantics implies that theoretical claims about this reality have truth values, and should be construed literally, whether true or false. ... Finally, the epistemological commitment is to the idea that these theoretical claims give us knowledge of the world. (2007: p9)

A1.3 Michael Devitt

Devitt wishes to state the claim of realism in explicitly metaphysical form:

> Most of the essential unobservables of well-established current scientific theories exist mind-independently and mostly have the properties attributed to them by science. (2004: p102)

He rejects the need for any reference to truth, claiming that 'no doctrine of truth is constitutive of metaphysical doctrines of realism' (*ibid:* p104). However, it is not that Devitt thinks that realist truth claims are wrong, merely that they are parasitic upon the core concern, which is the metaphysical claim that the *essential unobservables* actually do exist. Thus he thinks that the following is typical of the claims of others as to the nature of scientific realism:

> Most of the theoretical terms of currently well-established scientific theories refer to mind-independent entities and the theories' statements about those entities are approximately true. (*ibid:* p103)

Devitt thinks this can be seen as a paraphrase of his metaphysical statement in which 'refer' and 'true' are removed by the disquotational schema:

'F' refers iff Fs exist

'S' is true iff S

He stresses the importance of the metaphysical claim:

> Such paraphrases are often convenient but they do not change the subject matter away from atoms, viruses, photons, and the like. They are not in any interesting sense semantic. In particular they do not involve commitment to a causal theory of reference or a correspondence theory of truth, nor to any other theory of reference or truth. (*ibid:* p103)

It does seem plausible to focus on the metaphysical claim as representing the heart of scientific realism, though as we shall see, many other realists ignore his warnings and place their emphasis on semantic and epistemic claims such as correspondence truth and a causal theory of reference. Devitt can also be considered to be an *entity realist*, a position discussed below.

A1.4 James Ladyman

With his commitment to *Ontic Structural Realism*, Ladyman is not anti-realist in the sense that van Fraassen is (see, for example, his 2007, §3). Nevertheless, he is not a conventional realist either. He says that scientific realism is the view that

> we ought to believe that our best current scientific theories are approximately true, and that their central theoretical terms successfully refer to the unobservable entities they posit. (2007: p68)

This again avoids any explicit metaphysical commitment – here scientific realism is not the claim that theories are true, but the epistemic claim that we ought to believe that they are true and successfully refer – a semantic claim.

A1.5 Alan Musgrave

Musgrave is an interesting example of the range of views represented by the phrase 'scientific realist'. He is undoubtedly a realist, but of a very different stripe. So different, indeed, that his (1996) includes less space devoted to the defence of his realist position than is devoted to attacking the realism of others such as Boyd and Leplin! [1] Musgrave characterises realism as primarily

> a thesis about the aim of science. It says that the aim of a scientific inquiry is to discover the truth about the matter inquired into. This incorporates a semantic thesis (inquiry results in true or false statements about the world) and an axiological thesis (science aims for true statements). (1996: p19)

He also makes the correspondence theory of truth an integral part of his scientific realism, claiming that it is obvious that

> realism assumes some version of the classical or objective or correspondence (or we might add, realist) theory of truth. To say that the aim of inquiry is truth in any other sense of the term 'truth' is not to advocate realism at all. (*ibid:* p23).

---

[1] Musgrave relates that Bill Lycan called him a 'mad-dog realist' (1996: p19), though he thinks this title more correctly belongs to what he calls 'the substantive ontological or metaphysical thesis' which he attributes to Boyd and Leplin: 'that there really are such entities as current science claims there to be and that what current science tells us about such entities is true' (*ibid:* p20). He claims this kind of realism 'erects current science into a metaphysic and ties scientific realism too closely to that metaphysic', just as 'Cartesians did in the 17th century, Newtonians in the 18th century' (*ibid:* p21).

Musgrave is a fallibilist and entreats considerably more caution regarding the claims of scientific theories, a position much closer to mine in rejecting the more optimistic realism in which current theoretical claims are claimed to be mainly true:

> We should be more confident about atoms and molecules than we are about electrons, and more confident about electrons than we are about quarks and gluons. Realism about the entities and theories of current science should be rather guarded. And whether guarded or not, it should not be seen as definitional of scientific realism. (*ibid:* p22).

A1.6 Ilkka Niiniluoto

Niiniluoto says that what he calls 'critical scientific realism' comprises the following claims:

> R0 At least part of reality is ontologically independent of human minds.
>
> R1 Truth is a semantical relation between language and reality. Its meaning is given by a modern (Tarskian) version of the correspondence theory, and its best indicator is given by systematic enquiry using the methods of science.
>
> R2 The concepts of truth and falsity are in principle applicable to all linguistic products of scientific enquiry, including observation reports, laws, and theories. In particular, claims about the existence of theoretical entities have a truth value.
>
> R3 Truth (together with some other epistemic utilities) is an essential aim of science.

R4 Truth is not easily accessible or recognizable, and even our best theories can fail to be true. Nevertheless, it is possible to approach the truth, and to make rational assessments of such cognitive progress.

R5 The best explanation for the practical success of science is the assumption that scientific theories in fact are approximately true or sufficiently close to the truth in the relevant respects. Hence, it is rational to believe that the use of the self-corrective methods of science in the long run has been, and will be, progressive in the cognitive sense. (1999: p10)

We can set aside R5 since this cannot reasonably be counted as constitutive of scientific realism, but is, rather, the putative conclusion of the No Miracles Argument which I shall discuss in detail in §8. R0 is the metaphysical claim, R2 the semantic claim, R4 the epistemic claim, R3 an axiological claim, and R1 claims the correspondence theory of truth.

A1.7 Stathis Psillos

Psillos sets out these three theses which he thinks characterise scientific realism:

The Metaphysical Thesis: The world has a definite and mind-independent structure.

The Semantic Thesis: [Scientific theories] are capable of being true or false. The theoretical terms featuring in theories have putative factual reference. So, if scientific theories are true, the unobservable entities they posit populate the world.

Appendix 1

> *The Epistemic Thesis*: Mature and predictively successful scientific theories are well-confirmed and approximately true of the world. So, the entities posited by them, or, at any rate, entities very similar to those posited, inhabit the world. (2000: p707) [2]

He likes to associate the metaphysical part of his statement of scientific realism with a specific commitment to a *natural kinds* view of the world:

> [The first thesis] implies that if the unobservable natural kinds posited by theories exist at all, they exist independently of our ability to be in a position to know, verify, recognise etc. that they do.

He also goes on to associate realism with a non-epistemic view of truth:

> A non-epistemic account of the concept of truth is motivated to provide the best way to capture the intuition that scientific discourse is about a 'mind-independent' world, that is a world whose structure and content are logically and conceptually independent of the epistemic standards science uses to appraise theories. (1999: pxxi)

---

[2] Here we see a blatant misuse of the term *epistemic realism*. For Psillos doesn't say that we have warrant to believe what he claims, but that it is so – an ontological claim.

Later in the book it becomes clear that this is to be the correspondence theory:

> This section will defend a well-known argument in favour of a substantive 'correspondence' account of truth. (1999: p246)

Howard Sankey

Sankey (2008: pp12-17) presents six 'doctrines which form the core of scientific realism':

1 *Aim realism:* the aim of science is to discover the truth about the world, and scientific progress consists in advance toward that aim.

2 *Epistemic realism:* scientific inquiry leads to genuine knowledge of both observable and unobservable aspects of the world.

3 *Theoretical discourse realism:* scientific discourse about theoretical entities is to be interpreted in literal fashion as discourse that is genuinely committed to the existence of real unobservable entities.

4 *Metaphysical realism:* the world investigated by science is an objective reality that exists independently of human thought.

5 *Correspondence theory of truth*: truth consists in correspondence between a claim about the world and the way the world is.

6 *Objectivity of truth:* theories or claims about the world are made true (or false) by the way things are in the mind-independent, objective reality investigated by science.

According to Sankey, claim 6 is required only to ensure that claim 5 cannot be fitted into some kind of idealist version of claim 5. Clearly, Sankey's scientific realism is very wide in its claims.

A1.8 Wilfrid Sellars

Sellars formulates realism thus:

> ... to have good reason for holding a theory is *ipso facto* to have good reason for holding that the entities postulated by the theory exist. (1963: p97f)

However, these references to 'reason for holding' seem to lay stress on the epistemology. So here we have an epistemic thesis – that to hold a theory is equivalent to having good reason for thinking its theoretical terms exist.[3] However, of course Sellars also said, more famously:

> that in the dimension of describing and explaining the world, science is the measure of all things, of what is that it is, and of what is not that it is not. (*ibid:* p.173)

Clearly this is a metaphysical claim.

---

[3] A view strongly opposed by van Fraassen and the subject of a dispute between him and Sellars – see van Fraassen, 1975 and Sellars, 1976.

A1.9 Bas van Fraassen

van Fraassen says that:

> Science aims to give us, in its theories, a literally true story of what the world is like: and acceptance of a scientific theory involves the belief that it is true. This is the correct statement of scientific realism. (1980: p8)

van Fraassen here gives a definition that has nothing to do with any metaphysical commitments, but is instead a mixture of axiological ('science aims') and epistemic ('acceptance involves the belief...') criteria.

Appendix 2: The Base Rate Fallacy Argument

In recent years several writers have argued that both the NMA and the PI argument are flawed due to a problem known as the 'base rate' fallacy. This is one of the arguments in the realism/anti-realism debate that sprang up and quickly attracted several papers but has shown no conclusive result. However, my aim will be to show that it cannot tell against the argument as I have presented it.

I will first describe the argument as it is made against the NMA. Colin Howson (2000: §3), P.D. Magnus & Craig Callender (2004), and Peter Lipton (2004: §11) argue that the NMA is flawed because in order to evaluate the claim that theories enjoying empirical success are approximately true we have to know what the relevant base rate is, and there is no way we can know this. Lipton says:

> it is difficult not to suspect that the original plausibility of the miracle argument is just an instance of philosophers falling for the ubiquitous fallacy of ignoring base rates (2004, p196)

Lipton (2004) explains the base rate fallacy by way of a question set to students in the Harvard Medical School – a medical test example with the following data known: A test for a disease has a false negative rate of nil – i.e. nobody who has the disease will test negative. The false positive rate is 5% – five in every hundred people who do not have the disease will still test positive, and overall just 1 in 1000 people have the disease. Now given these known facts, if a specific patient tests positive, what then is the probability that this patient has the disease? Most gave the answer 0.95 but the correct answer is just under 0.02 for this reason: If the test is given to 1000 people, about 50 will test positive even though they don't have the disease (false positive rate of 5%), plus the 1 who tests positive because she does have it, so that makes an expected 51 people will test positive. But of those, only 1 really has the disease, so the probability that this specific patient has the disease is 1 in 51 – less than .02. The high proportion of people who get this wrong is said to demonstrate the ubiquity of the base rate fallacy and Lipton suggests that the NMA seems to work the same way, 'with realists playing the role of the probabilistically challenged medics' (ibid: p197). The correspondence is:

- Being true is the disease.
- Making lots of successful predictions is testing positive for it.
- The false negative rate is nil: no true theory makes false predictions.
- The false positive rate is low, since relatively few false theories are empirically successful.

Thus we are inclined to infer that highly successful theories are likely to be true. What we ignore is the base rate – that the vast majority of theories are false, so even a very small probability that a false theory should make such successful predictions leaves it the case that the great majority of successful theories are false. Most false theories are unsuccessful, but alas what counts is that most successful theories are false.

Lipton asks why so many of us are prone to commit the base rate fallacy here and says we focus on the comparison between the relatively few false theories that are successful and the many false theories that are unsuccessful, while ignoring the comparison between the very few true theories and the many more false theories that are successful. Thus when Putnam memorably says that it would take a miracle for a false theory to make all those true predictions, this way of putting it directs our attention to the very low proportion of successful theories among false theories, but this has the effect of diverting our attention away from the low proportion of true theories among successful theories that is more to the point. (*ibid:* p197)

Psillos (2006: §7) argues against this 'subjective Bayesian' argument on a number of grounds, including questioning the argument's claim that the realist has failed to take into account the base rates for true or false successful theories. Psillos claims there are philosophical grounds for doubting that there are any such base rates:

> The very idea of a base-rate of truth and falsity depends on how the *relevant* population of theories is fixed. This is where many philosophical problems loom large. For one, we don't know how exactly we should individuate and count theories. For another, we don't even have, strictly speaking, outright true and false theories. But suppose that we leave all this to one side. A more intractable problem concerns the concept of success. What is it for a theory to be successful? (2006: p21)

These seem reasonable points, though I might point out that the last two sentences tell against the whole NMA. Psillos also claims that 'a probabilistic argument is *deeply* problematic: it fails to capture the rich structure of theory-change in science.' (*ibid:* p21)

There are similar base rate arguments against the PMI (Lewis, 2001 and Lange, 2002) and once again there are counter-arguments that say the PMI is untouched (e.g. Saatsi, 2005). Magnus and Callender (2004) argue that the applicability of the base rate argument to both NMA and PI should signal the dissolution of the whole debate. So we have:

Base rate fallacy refutes NMA:

> For: Howson, Lipton, Magnus & Callender.
> Contra: Psillos.

Base rate fallacy refutes PI:

> For: Lewis, Lange, Magnus & Callender.
> Contra: Saatsi.

I am sceptical that the debate between NMA and PI is fundamentally about statistics and induction. I don't think the NMA is fundamentally an induction at all but is based on the claim that one is entitled to infer approximate truth of a theory from its predictive success. It is true that the NMA starts out referring to the 'success of science' – i.e. to a whole population of theories, but it rapidly moves to focussing on specific theories. However, even if it were conceded to have some force, this would merely add to the arguments against NMA and thus strengthen my case. Regarding the PI, I showed in §5 that there are even stronger reasons to set aside inductive issues in favour of viewing the PI as a direct evidential refutation of the move from success to approximate truth. So in both cases considerations of probabilities and base rates would be irrelevant.

Appendix 2

It is sufficient if the PI succeeds in demonstrating the existence of just one highly successful theory that is clearly neither true nor approximately true. For if it does then the explanation of the success of *that* theory cannot be approximate truth but must be something else, and whatever that may be we can assume that it might be the explanation of the success of *all* theories. The whole basis of the NMA is thus defeated. Now in fact Laudan's list shows several such theories and neither the move to mature theories (§6.4) and novel predictive success (§6.5), nor the selective truth realism appeal to idle wheels (§6.6) removes *all* of these – caloric and ether theories, for example. This logic is clearly not inductive. The realist may argue that one successful theory which is not approximately true is not enough to upset her argument, that this might have been a coincidence. However, serious consideration of the caloric and material ether theories with their richness and complexity makes this idea absurd – it really *would* be a miracle if the predictive success of these theories were just a coincidence or fluke. If there is even just *one* such theory, then the inference from success to approximate truth is not simply made less plausible but is shown to be wholly unjustified.

Consequently, I am entitled to say that the base rate fallacy argument is not relevant to my argument, for this reason:

a) If the base rate fallacy does not tell against NMA then it has no affect on my argument.

b) If the base rate fallacy does tell against NMA then that strengthens my argument.

c) If the base rate fallacy does not tell against PI then it has no affect on my argument.

d) The base rate fallacy *cannot* tell against PI in the deductive form I presented.

So the base rate argument either has no effect on my argument against the NMA or strengthens it. On the other hand, it cannot affect my presentation of the PI in its deductive form. Consequently, whatever conclusion is eventually reached in the base rates debate cannot affect the argument of this essay.

References

Achinstein, P. (2002), 'Is There a Valid Experimental Argument for Scientific Realism?', *Journal of Philosophy* 99: 470-495.

Agassi, J. (1960), 'Methodological Individualism', *British Journal of Sociology*, II, 1960, pp244-270, and in J. O'Neill (ed.), *Modes of Individualism and Collectivism* (London: Heinemann, 1973).

Annas, J. & Barnes, J. (eds.) (1985), *The Modes of Scepticism: Ancient Texts and Modern Interpretations* (Cambridge: Cambridge University Press)

Armstrong, D. (1991), 'The Causal Theory of Mind' in D. Rosenthal (ed.) *The Nature of Mind* (NY, Oxford University Press).

Aronson, J. L. (1990), 'Verisimilitude and Type Hierarchies', *Philosophical Topics* 18: 5-28.

Aronson, J. L. & Harré, R. & Way, E. C. (1994), *Realism Rescued: How Scientific Progress is Possible* (London: Duckworth).

Barnes, B. & Bloor, D. (1982), 'Relativism, Rationalism and the Sociology of Knowledge', in M. Hollis and S. Lukes (eds.) *Rationality and Relativism* (Oxford: Basil Blackwell, 1982).

Bird, A. (2000), *Thomas Kuhn* (Chesham: Acumen).

Bird, A. (2007a), *Nature's Metaphysics: Laws and Properties* (Oxford: Oxford University Press)

Bird, A. (2007b), 'What is Scientific Progress?', *Noûs* 41:1 (2007) pp64-89.

Bird, A. (2010), 'The Epistemology of Science – a Bird's-Eye View', *Synthese* (forthcoming).

Bishop, M. (2003), 'The Pessimistic Induction, the Flight to Reference, and the Metaphysical Zoo', *International Studies in the Philosophy of Science*, vol. 17, No. 2, 2003.

Bishop, M. & Stich, S. (1997), 'The Flight to Reference or How Not to Make Progress in the Philosophy of Science', *Philosophy of Science*, 65 (March 1998) pp33-49.

Blackburn, S. (2002), 'Realism: Deconstructing the Debate', *Ratio* 15: 111–133.

Blake, R.M. (1960), 'Theory of Hypothesis Among Renaissance Astronomers', in R.M. Blake (ed.) *Theories of Scientific Method* (Seattle: University of Washington Press).

Bloor, B. (1991), *Knowledge and Social Imagery* (Chicago: University of Chicago Press)

Boltzmann, L. (1905), 'Theories as representations', in A. Danto and S. Morgenbesser (eds.) *Philosophy of Science*, pp245-252, (New York: World Publishing Co., 1960).

Boyd, R. (1973), 'Realism, Underdetermination, and a Causal Theory of Evidence', *Noûs* 7: 1-12.

Boyd, R. (1983), 'On the Current Status of the Issue of Scientific Realism' in *Erkenntnis* 19, p45-90.

Boyd, R. (1984), 'The Current Status of Scientific Realism', in Leplin, 1984.

Boyd, R. (1989), 'What Realism Implies and What it Does Not' in *Dialectica* Vol. 43, No. 1-2 (1989), pp5-29.

Boyd, R. (2002), 'Scientific Realism' in *The Stanford Encyclopaedia of Philosophy* (Fall 2008 Edition), Edward N. Zalta (ed.), URL = <http://plato.stanford.edu/archives/win2003/entries/scientific-realism/>.

Bueno, O. (1999), 'What is Structural Empiricism? Scientific Change in an Empiricist Setting', *Erkenntnis* 50: 59–85.

Bueno, O. (2000), 'Empiricism, Scientific Change and Mathematical Change', *Studies in History and Philosophy of Science* 31: 269–296.

Carnap, R. (1956), 'The Methodological Character of Theoretical Concepts' in *Minnesota Studies in the Philosophy of Science*, vol. I, ed. H. Feigl and M. Scriven (Minneapolis : University of Minnesota Press).

Carrier, M. (1991), 'What is wrong with the miracle argument?' *Studies in History and Philosophy of Science*, 22(1), 23–36.

Cartwright, N. (1983), *How the Laws of Physics Lie* (Oxford: Clarendon).

Cartwright, N. (1999), *The Dappled World: A Study of the Boundaries of Science* (Cambridge: Cambridge University Press).

Chakravartty, A (2007), *A Metaphysics for Scientific Realism* (Cambridge: Cambridge University Press)

Chakravartty, A (2008), 'What You Don't Know Can't Hurt You: Realism and the Unconceived', *Philosophical Studies* (2008) 137:pp149–158.

Chang, H. (2003a), *Inventing Temperature: Measurement and Scientific Progress* (Oxford: Oxford University Press).

Chang, H. (2003b), 'Preservative Realism and Its Discontents: Revisiting Caloric', *Philosophy of Science* 70: 902-912.

Chang, H. (forthcoming), *Is Water H2O? Evidence, Realism, Pluralism* (Boston Studies in the Philosophy of Science, Springer).

Churchland, P. (1985), 'The Ontological Status of Observables: In Praise of the Superempirical Virtues', in Churchland & Hooker (1985).

Churchland, P. & Hooker, C. (eds.) (1985), *Images of Science, Essays on Realism and Empiricism, with a Reply by Bas C. van Fraassen* (Chicago: Chicago University Press)

Clarke, S. & Lyons, T. (2002), 'Introduction: Scientific Realism and Commonsense', in S. Clarke & T. Lyons (eds.) *Recent Themes in the Philosophy of Science: Scientific Realism and Common Sense* (Dordrecht: Kluwer).

Collins, H. & Pinch, T. (1998), *The Golem: What You Should Know about Science*, 2nd edn. (Cambridge: Cambridge University Press).

Cornman, J. (1976), 'Sellars on Scientific Realism and Perceiving', *Proceedings of the Biennial Meeting of the PSA*, 1976 vol. 2: *Symposia and Invited Papers.*

Cushing, T. & Delaney, C. & Gutting, G. (eds.) (1983), *Science and Reality: Recent Work in the Philosophy of Science* (Indiana: University of Notre Dame Press).

da Costa, Newton C.A. & French, S. (2003), *Science and Partial Truth, A Unitary Approach to Models and Scientific Reasoning* (Oxford: Oxford University Press)

Darling, K. M. (2002), 'The complete Duhemian Underdetermination Argument: Scientific Language and Practice', *Studies in History and Philosophy of Science*, 33 (2002) 511-533.

Davidson, D (1973), 'Radical Interpretation', *Dialectica*, 27, reprinted in his *Inquiries into Truth and Interpretation* (Oxford: Clarendon Press).

De Regt, H. (1994), Representing the World by Scientific Theories: The Case for Scientific Realism (Tilburg: Tilburg University Press).

Devitt, M. (1997), *Realism and Truth*, 2nd edn. (Princeton: Princeton University Press).

Devitt, M. (2002), 'Underdetermination and Realism', in E. Sosa & E. Villaneuva (eds.), *Realism and Relativism, Philosophical Issues Vol. 12* (Oxford: Blackwell).

Devitt, M. (2004), 'Scientific Realism', in P. Greenough & M. Lynch (eds.), *Truth and Realism* (Oxford: Oxford University Press), also in F. Jackson & M. Smith (eds.) *The Oxford Handbook of Contemporary Philosophy* (Oxford: Oxford University Press, 2005).

Devitt, M. (2010), *Putting Metaphysics First: Essays on Metaphysics and Epistemology* (Oxford: Oxford University Press).

Douven, I. (2008), 'Underdetermination', in S. Psillos & M. Curd (eds.), *The Routledge Companion to Philosophy of Science*, pp292-301 (Abingdon: Routledge).

Duhem, P (1908), *To Save the Phenomena*, trans. E. Doland & C. Mascher (Chicago: Chicago University Press, 1969).

Duhem, P (1914), *The Aim and Structure of Physical Theory* (Princeton University Press, 1991).

Dupre, J. (1993), *The Disorder of Things: Metaphysical Foundations of the Disunity of Science* (Harvard: Harvard University Press).

Earman, J. (1993), 'Underdetermination, Realism and Reason', *Midwest Studies in Philosophy* 18: 19-38.

Ellis, B. (1979), *Rational Belief Systems* (Oxford: Blackwell).

Ellis, B. (1985), 'What Science Aims to Do', in Churchland and Hooker (1985).

Ellis, B. (1990), *Truth and Objectivity* (Oxford: Blackwell).

Ellis, B. (2001), *Scientific Essentialism* (Cambridge: Cambridge University Press).

Ellis, B. (2002), *The Philosophy of Nature: A Guide to the New Essentialism* (Chesham: Acumen).

Ellis, B. (2005), 'Physical Realism', *Ratio*, XVIII, 4 Dec 2005.

Elsamahi, M. (2005), 'A Critique of Localized Realism', *Philosophy of Science* 72: 1350-1360.

Elster, J. (1982), 'The Case for Methodological Individualism', *Theory and Society*, 11: pp453–482.

Evans, G. (1973), 'The Causal Theory of Names', *Proceedings of the Aristotelian Society*, 47, 187-208, reprinted in Schwartz, S. (ed.) *Naming, Necessity and Natural Kinds* (Ithaca: Cornell University Press, 1977), pp192-215.

Feyerabend, P. (1975), *Against Method* (London: Verso).

Field, H. (1973), 'Theory-Change and the Indeterminacy of Reference', *Journal of Philosophy*, 70, 462-81.

Fine, A. (1984), 'The Natural Ontological Attitude' in Leplin, 1984.

Fine, A. (1986a), 'Unnatural Attitudes: Realist and Instrumentalist Attachments to Science', *Mind*, 95: 149-179.

Fine, A. (1986b), *The Shaky Game* (Chicago, University of Chicago Press).

Fine, A. (1991), 'Piecemeal Realism' in *Philosophical Studies*, Feb 1991, pp79-96.

French, S. (2003), 'Scribbling on the Blank Sheet: Eddington's Structuralist Conception of Objects', *Studies in History and Philosophy of Modern Physics* 34: 227–259.

French, S. & Ladyman, J. (2003), 'Remodelling Structural Realism: Quantum Physics and the Metaphysics of Structure', *Synthese* 136: 31-56.

French, S. & Ladyman, J. (forthcoming), 'In Defence of Ontic Structural Realism' in Alissa and Bokulich (eds.) *Boston Studies in the Philosophy of Science: Scientific Structuralism* (forthcoming from Boston University Press).

Frost-Arnold, G. (2010), 'The No-Miracles Argument for Realism: Inference to an Unacceptable Explanation', in *Philosophy of Science*, 77/1, January 2010: 35-58.

Fuller, S. (2000), *Thomas Kuhn: A Philosophical History for our Times* (Chicago: University of Chicago Press).

Fuller, S. (2003), *Kuhn vs. Popper: The Struggle for the Soul of Science* (Cambridge: Icon Books).

Giere, R. (1988), *Explaining Science, A Cognitive Approach* (Chicago: University of Chicago Press).

Giere, R. (2006), 'Perspectival Pluralism', in Kellert, Longino & Waters, 2006).

Glymour, C. (1971), 'Theoretical Realism and Theoretical Equivalence', *Boston Studies in the Philosophy of Science*, vol. 8 (Dordrecht: Reidel Publishing Company).

Glymour, C. (1980), *Theory and Evidence*, (Princeton: Princeton University Press).

Gower, B. (2000), 'Cassirer, Schlick and Structural Realism: the Philosophy of the Exact Sciences in the Background to Early Logical Empiricism', *British Journal for the History of Philosophy* 8: 71–106.

Grandy, R. (1973), 'Reference, Meaning, and Belief', *Journal of Philosophy*, 70, 439-452.

Hacker, P.M.S. (1987), *Appearance and Reality* (Oxford: Blackwell).

References

Hacking, I. (1983), *Representing and Intervening* (Cambridge, Cambridge University Press).

Hacking, I. (1996), 'The Disunities of the Sciences' in P. Galison & D. Stump (eds.) *The Disunity of Science, Boundaries, Contexts, and Power* (Stanford: Stanford University Press).

Harman, P.M. (1998), *The Natural Philosophy of James Clerk Maxwell* (Cambridge: Cambridge University Press).

Hardin, C. & Rosenberg, A. (1982), 'In Defense of Convergent Realism', *Philosophy of Science*, Vol. 49, No. 4. (Dec., 1982), pp. 604-615.

Hempel, C. G. (1965), *Aspects of Scientific Explanation* (New York: Free Press).

Hilpinen, R. (1976), 'Approximate Truth and Truthlikeness', in M. Przelecki, K. Szaniawski, & R. Wojcicki (eds.), *Formal Methods in the Methodology of Empirical Sciences* (Dordrecht: Reidel Publishing Company).

Hertz, H. (1893), *Electric Waves*, tr. D. E. Jones (New York: Dover, 1962).

Hertz, H. (1900), *The Principles of Mechanics Presented in a New Form* (New York: Dover, 2003)

Hoefer, C. & Rosenberg, A. (1994), 'Empirical Equivalence, Underdetermination, and Systems of the World', *Philosophy of Science* 61: 592-608.

Horwich, P. (1982a), 'How to Choose Between Empirically Indistinguishable Theories', *Journal of philosophy* 79: 61-77.

Horwich, P. (1982b), *Probability and Evidence* (Cambridge: Cambridge University Press).

Horwich, P. (1991), 'On the Nature and Norms of Theoretical Commitment', *Philosophy of Science* 58: 1-14.

Howson, C. (2000), *Hume's Problem: Induction and the Justification of Belief* (Oxford: Oxford University Press).

Jackson, F. (1991), "What Mary Didn't Know", in D. Rosenthal (ed.) *The Nature of Mind* (NY, Oxford University Press).

Jackson, F. (2003), 'Mind and Illusion', in O'Hear, A. (ed.) *Minds and Persons* (Cambridge: Cambridge University Press).

Kekulé, A. (1867), 'On Some Points of Chemical Philosophy', *The Laboratory*, I, July 27, 1867. Reprinted in R. Anschütz, *August Kekulé*, vol. 2, Berlin, 1929.

Kellert, H. & Longino, H. & Waters, C. (eds.) (2006) *Scientific Pluralism, Minnesota Studies in the Philosophy of Science* vol. 19 (Minneapolis: University of Minnesota Press).

Kitcher, P. (1993), *The Advancement of Science* (Oxford: Oxford University Press).

Kitcher, P. (2001a), 'Real Realism: The Galilean Strategy', *The Philosophical Review* 110, pp. 151-197.

Kitcher, P. (2001b), *Science, Truth, and Democracy* (Oxford: Oxford University Press).

Krajewski, W. (1977), *Correspondence Principle and Growth of Science*, (Dordrecht: Reidel Publishing Company).

Kripke, S. (1972), *Naming and Necessity* (Harvard: Harvard University Press).

Kuhn, T. (1957), *The Copernican Revolution* (Cambridge, Mass.: Harvard University Press).

Kuhn, T. (1996), *The Structure of Scientific Revolutions* 3rd ed. (Chicago: Chicago University Press).

Kuipers, T. (ed.) (1987) *What Is Closer-to-the-Truth?* (Amsterdam, Rodopi).

Kukla, A. (1993), 'Laudan, Leplin, Empirical Equivalence and Underdetermination', *Analysis* 53: 1-7.

Kukla, A. (1994), 'Non-Empirical Theoretical Virtues and the Argument from Underdetermination', *Erkenntnis* 41: 157-170.

Kukla, A. (1996), 'Does Every Theory Have Empirically Equivalent Rivals?', *Erkenntnis* 44: 137-166.

Kukla, A. (1998), *Studies in Scientific Realism* (Oxford: Oxford University Press).

Ladyman, J. (1998), 'What is structural realism?'. *Studies in History and Philosophy of Science* 29: 409–424.

Ladyman, J. (2000), 'What's Really Wrong with Constructive Empiricism? van Fraassen and the Metaphysics of Modality', *The British Journal for the Philosophy of Science*, 51, pp837-56.

Ladyman, J. (2002), 'Science, Metaphysics and Structural Realism', *Philosophica*, 67: 57-76.

Ladyman, J. (2004), 'Constructive Empiricism and Modal Metaphysics: a Reply to Monton and van Fraassen', *The British Journal for the Philosophy of Science*, 55, pp755-65.

Ladyman, J. (2006), 'Does Physics Answer Metaphysical Questions', in A. O'Hear (ed.), *Philosophy of Science: Royal Institute of Philosophy Supplement 61* (Cambridge, Cambridge University Press).

Ladyman, J. & Ross, D. (2007), *Everything Must Go, Metaphysics Naturalised* (Oxford: Oxford University Press).

Ladyman, J. & Douven, I. & Horsten, L. & van Fraassen, B. (1997), 'A Defence of van Fraassen's Critique of Abductive Inference: Reply to Psillos', *Philosophical Quarterly*, vol. 47, No. 188, pp305-321.

Latour, B. & Woolgar, S. (1979), *Laboratory Life: The Social Construction of Scientific Facts* (London: Sage).

Lakatos, I. (1970), 'Falsification and the Methodology of Research Programmes' in I. Lakatos & A. Musgrave (eds.) *Criticism and the Growth of Knowledge* (Cambridge: Cambridge University Press).

Laudan, L. (1977), *Progress and its Problems: Towards a Theory of Scientific Growth* (London: Routledge & Kegan Paul).

Laudan, L. (1981), 'A Confutation of Convergent Realism', *Philosophy of Science*, vol. 48, no. 1 (Mar 1981) pp19-49, reprinted in Leplin, 1984.

Laudan, L. (1983), 'Explaining the Success of Science', in Cushing, Delaney, and Gutting, 1983.

Laudan, L. (1984a), 'Realism without the Real', *Philosophy of Science*, vol. 51, no. 1 (Mar 1984) pp156-162.

Laudan, L. (1984b), *Science and Values* (Berkeley: University of California Press).

Laudan, L. (1985), 'Realism without the Real', *Philosophy of Science*, vol. 51, no. 1 (Mar 1984) pp156-162.

Laudan, L. (1990a), *Science and Relativism* (Chicago: University of Chicago Press).

Laudan, L. (1990b), 'Demystifying Underdetermination', in C.W. Savage (ed.) *Scientific Theories: Minnesota Studies in the Philosophy of Science* vol. 14, (Minneapolis: University of Minnesota Press).

Laudan, L. (1996), *Beyond Positivism and Relativism* (Boulder: Westview Press).

Laudan, L. & Leplin, J. (1991), 'Empirical Equivalence and Underdetermination', *Journal of Philosophy* 88: 449-472.

Leeds, S. (1994), 'Constructive Empiricism', *Synthese*, 101 (1994), pp187–221.

Leplin, J. (ed.) (1984a), *Scientific Realism* (London: University of California Press).

Leplin, J. (1984b), 'Truth and Scientific Progress', in his (1984a).

Leplin, J. (1987), 'Surrealism', *Mind* 96: pp519-524.

Leplin, J. (1997), A Novel Defense of Scientific Realism (Oxford: Oxford University Press).

Lewis, P. (2001), 'Why the Pessimistic Induction is a Fallacy', *Synthese* 129: 371-380.

Lipton, P. (1994), 'Truth, Existence and the Best Explanation', in A. A. Derksen (ed.) *The Scientific Realism of Rom Harré* (Tillburg: Tillburg University Press), pp. 89-111.

Lipton, P. (2004), *Inference to the Best Explanation*, 2nd edn. (London: Routledge).

Loewer, B. (1996), 'Humean Supervenience' in *Philosophical Topics* 24:101-123.

Lyons, T. (2002), 'The Pessimistic Meta-Modus Tollens', in S. Clarke & T. Lyons (eds.) *Recent Themes in the Philosophy of Science: Scientific Realism and Common Sense* (Dordrecht: Kluwer).

Lyons, T. (2003), 'Explaining the Success of a Scientific Theory', *Philosophy of Science*, 70(5) (2003), pp891-901.

Lyons, T. (2005), 'Toward a Purely Axiological Scientific Realism', *Erkenntnis*, 63:2005, pp167-204.

Lyons, T. (2006), 'Scientific Realism and the Stratagema de Divide et Impera', *British Journal for Philosophy of Science*, 57 (2006), pp537-60.

Magnus, P. D. & Callender, C. (2004), 'Realist Ennui and the Base Rate Fallacy', *Philosophy of Science* 71, July 2004, pp. 320-338.

Mach, E. (1883), *The Science of Mechanics*, tr. T. McCormack (1893) (Illinois: Open Court, 1974).

Mach, E. (1885), *Contributions to the Analysis of the Sensations*, tr. C.M. Williams (1896) (Illinois: Open Court).

Matheson, C. (1998), 'Why the No Miracles Argument Fails', *International studies in the philosophy of science*, vol. 12, no. 3.

Maxwell, J.C. (1873), *A Treatise on Electricity and Magnetism*, 2 vols. (London: Oxford University Press, 1955).

Maxwell, J. C. (1881), *Elementary Treatise on Electricity* (London: Dover Publications, 2005).

Maxwell, G. (1962), 'The Ontological Status of Theoretical Entities', in H. Feigl and G. Maxwell (eds.) *Scientific Explanation, Space and Time: Minnesota Studies in the Philosophy of Science vol. 3* (Minneapolis: University of Minnesota Press).

McAllister, J. (1993), 'Scientific Realism and the Criteria for Theory-Choice', *Erkenntnis*, 38, 203-222.

McGinn, C. (1995), 'Consciousness and Space', in *Journal of Consciousness Studies* 2, 221-230.

McLeish, C. (2005), 'Scientific Realism Bit by Bit: Part I, Kitcher on Reference', *Studies in History and Philosophy of Science*, 36(2005), 667-685.

McMullin, E. (1984), 'A Case for Scientific Realism' in Leplin (1984a).

McMullin, E. (1991), 'Selective Anti-realism', *Philosophical Studies*, 61 (1991), pp97-108.

Mill, J. S. (1843), *System of Logic* (Longmans, 1961).

Miller, D. (1974), "Popper's Qualitative Theory of Verisimilitude", *The British Journal for the Philosophy of Science*, 25: 166-177.

Miller, D. (1976), 'Verisimilitude Redeflated', *British Journal for the Philosophy of Science* 27: 363-380.

Misak, C.J. (2004), *Truth and the End of Inquiry* (Oxford: Oxford University Press).

Monton, B. (ed.) (2007), *Images of Empiricism: Essays on Science and Stances, With a Reply from Bas van Fraassen* (Oxford: Oxford University Press).

Monton, B. & Mohler, C. (2008), 'Constructive Empiricism', *The Stanford Encyclopedia of Philosophy (Winter 2008 Edition)*, Edward N. Zalta (ed.), URL = <http://plato.stanford.edu/archives/win2008/entries/constructive-empiricism/>.

Monton, B. & van Fraassen, B. (2003), 'Constructive Empiricism and Modal Nominalism', *The British Journal for the Philosophy of Science*, 54, pp405-22.

Mumford, S. (2004), *Laws in Nature* (Routledge).

Musgrave, A. (1985), 'Realism versus Constructive Empiricism', in Churchland & Hooker (1985).

Musgrave, A. (1988), 'The Ultimate Argument for Scientific Realism', in R. Nola (ed.) *Relativism and Realism in Science* (Boston: Kluwer).

Musgrave, A. (1996), 'Realism, Truth and Objectivity' in R. S. Cohen et al. (eds.), *Realism and Anti-Realism in the Philosophy of Science* (Dordrecht: Kluwer).

Musgrave, A. (2006), 'The 'Miracle Argument' For Scientific Realism' in *The Rutherford Journal* vol. 2, 2006-7, at www.rutherfordjournal.org.

Newton-Smith, W. (1978), 'The Underdetermination of Theories by Data', *Proceedings of the Aristotelian Society*: 71-91.

Niiniluoto, I. (1987), *Truthlikeness* (Dordrecht: Reidel Publishing Company).

Niiniluoto, I. (1998), 'Verisimilitude: The Third Period', *British Journal for the Philosophy of Science* 49: 1-29.

Niiniluoto, I (1999), *Critical Scientific Realism* (Oxford: Oxford University Press).

Nola, R. (1980), 'Fixing the Reference of Theoretical Terms', *Philosophy of Science*, Vol. 47, No. 4. (Dec., 1980), pp. 505-531.

Nye, M.J. (1972), *Molecular Reality. A Perspective on the Scientific Work of Jean Perrin* (American Elsevier).

O'Leary-Hawthorne, J. (1994), 'What Does Van Fraassen's Critique of Scientific Realism Show?', *The Monist*, Vol. 77, No. 1., pp128-145.

Oddie, G. (1986), *Likeness to Truth* (Dordrecht: Reidel Publishing Company).

Oderberg, D. (2007), *Real Essentialism* (Abingdon: Routledge).

Papineau, D. (ed.) (1996), *The Philosophy of science* (Oxford: Oxford University Press).

Poincaré, H. (1905), *Science and Hypothesis* (New York: Dover, 1952).

Poincaré, H. (1906), *The Value of Science*, tr. G.B. Halsted, 1914 (New York: Dover, 1958).

Popper, R. (1956), 'Three Views Concerning Knowledge', in (Popper: 1963).

Popper, R. (1963), *Conjectures and Refutations* (London: Routledge & Kegan Paul).

Popper, R. (1972), *Objective Knowledge* (Oxford: Oxford University Press).

Popper, R. (1983), Realism and the Aim of Science from The Postscript to the Logic of Scientific Discovery, W. W. Bartley III (ed.) (London: Routledge).

Preston, J.P. (2008), 'Hertz, Wittgenstein and Philosophical Method', *Philosophical Investigations* 31:1 January 2008.

Psillos, S. (1996), 'Scientific Realism and the Pessimistic Induction', *Philosophy of Science*, 63 (Proceedings) pp S306-S314.

Psillos, S. (1997), 'Kitcher on Reference', *Studies in History and Philosophy of Science*, 11(3), 259-272.

Psillos, S. (1999), Scientific Realism: How Science Tracks Truth (Abingdon: Routledge).

Psillos, S. (2000), 'The Present State of the Scientific Realism Debate', *British Journal for the Philosophy of Science*, vol. 51 (Supp): 705-728.

Psillos, S. (2006), 'Thinking about the ultimate argument for realism', in *Rationality and Reality: Conversations with Alan Musgrave* (Dordrecht: Springer).

Psillos, S. (2009), *Knowing the Structure of Nature* (London: Palgrave Macmillan).

Putnam, H. (1975a), 'What is Mathematical Truth?' in his *Mathematics, Matter, and Method : Philosophical Papers Vol. I* (Cambridge: Cambridge University Press).

Putnam, H. (1975b), 'Language and Reality' in his *Mind, Language and Reality: Philosophical Papers: Vol. II*, (Cambridge: Cambridge University Press).

Putnam, H. (1975c), 'How Not to Talk about Meaning', in his *Mind, Language and Reality: Philosophical Papers: Vol. II*, (Cambridge: Cambridge University Press).

Putnam, H. (1978), *Meaning and the Moral Sciences* (London: Routledge).

Putnam, H (1981), *Reason, Truth and History* (Cambridge: Cambridge University Press).

Putnam, H. (1987), 'Realism With a Human Face', Lecture Stanford, in his (1990).

Putnam, H. (1990), *Realism With a Human Face* (London: Harvard University Press).

Quine, W.V. (1975), 'On Empirically Equivalent Systems of the World', *Erkenntnis* 9: 313-328.

Quinn, P. L. (1974), 'What Duhem really meant', R. S. Cohen, & M. W. Wartofsky (eds.), *Methodological and Historical Essays in the Natural and Social Sciences* (Boston: Reidel).

Redhead, M. (2001), 'Quests of a Realist', Review article of Stathis Psillos's Scientific Realism: How Science Tracks Truth, *Metascience*, vol. 10(3): 341-347.

Reisch, G. (1991), 'Did Kuhn Kill Logical Empiricism?' *Philosophy of Science*, 58/2 (1991), pp264-277.

Rescher, N. (1982), *Empirical Inquiry* (Rowman & Littlefield Publishers).

Richards, R. J (1981), 'Natural Selection and Other Models in the Historiography of Science', in Brewer, B. & Collins, B. (eds.) *Scientific Inquiry and the Social Sciences* (San Francisco: Jossey-Bass).

Rosen, G. (1994), 'What is Constructive Empiricism', *Philosophical Studies* 74: 143-178.

Russell, B. (1912a), *The Problems of Philosophy* (Oxford: Oxford University Press).

Russell, B. (1912b), 'On the Notion of Cause', in J. Slater, ed., *The Collected Papers of Bertrand Russell v6: Logical and Philosophical Papers 1909-1913* (London: Routledge Press, 1992).

Ryckman, T. (2005), *The Reign of Relativity: Philosophy in Physics 1915–1925* (Oxford: Oxford University Press).

Saatsi, J. (2005), 'On the Pessimistic Induction and Two Fallacies', *Philosophy of Science*, 72 (December 2005) pp. 1088–1098.

Sankey, H. (2001), 'Scientific Realism: An Elaboration and a Defense', *Theoria*, 98, pp35-54.

Sankey, H. (2002), 'Realism, Method and Truth', in M. Marsonet (ed.), *The Problem of Realism* (Aldershot: Ashgate), pp. 64-81.

Sankey, H. (2008), *Scientific Realism and the Rationality of Science* (Aldershot: Ashgate Publishing Ltd.).

Schilpp, P. (ed.) (1974), The Philosophy of Karl Popper: Library of Living Philosophers, vol.14 (Illinois: Open Court).

Sellars, W. (1956), 'Empiricism and the Philosophy of Mind', in *Minnesota Studies in the Philosophy of Science*, vol. I, ed. H. Feigl and M. Scriven (Minneapolis: University of Minnesota Press), reprinted in his (1963).

Sellars, W. (1960), 'Philosophy and the Scientific Image of Man', in *Frontiers of Science and Philosophy*, ed. R. Colodny (Pittsburgh: University of Pittsburgh Press, 1962), reprinted in his (1963).

Sellars, W. (1963), *Science, Perception, and Reality* (New York: Humanities Press).

Sellars, W. (1976), 'Is Scientific Realism Tenable?', *PSA 1976*, Vol. 2., ed. F. Suppe and P.D. Asquith, pp307-334.

Skidelsky, E. (2008), *Ernst Cassirer, the Last Philosopher of Culture* (Oxford: Princeton University Press).

Sklar, L. (1975), 'Methodological Conservatism', *Philosophical Review* 84: 384-400.

Sklar, L. (1981), 'Do Unborn Hypotheses Have Rights?', *Pacific Philosophical Quarterly* 62: 17-29.

Smart, J. J. C. (1963), *Philosophy and Scientific Realism* (London: Routledge & Kegan Paul).

Smart, J. J. C. (1968), *Between Science and Philosophy* (New York: Random House).

Sparkes, A.W. (1991), *Talking Philosophy, a Wordbook* (London: Routledge).

Stachel, J. (2000), '"What Song the Syrens Sang": How Did Einstein Discover Special Relativity?', in his *Einstein from "B" to "Z"* (Basel: Birkhauser Verlag). See also http://www.aip.org/history/einstein/essay-einstein-relativity.htm.

Stanford, P.K. (2000), 'An Antirealist Explanation of the Success of Science', *Philosophy of Science*, Vol. 67, No. 2. (June, 2000), pp. 266-284.

Stanford, P.K. (2001), 'Refusing the Devil's Bargain: What Kind of Underdetermination Should We Take Seriously', *Philosophy of Science*, Vol. 68, No. 3 *Supplement: Proceedings of the 2000 Biennial Meeting of the Philosophy of Science Association. Part I: Contributed Papers, (Sep., 2001)*, pp. S1-S12.

Stanford, P.K. (2003), 'No refuge for Realism: Selective Confirmation and the History of Science', *Philosophy of Science*, Vol. 70 (2003) pp. 913–925.

Stanford, P.K. (2005), 'Instrumentalism', in S. Sarkar & J. Pfeifer (eds.), *The Philosophy of Science, an Encyclopaedia*, (New York: Routledge).

Stanford, P.K. (2006), *Exceeding Our Grasp, Science, History, and the Problem of Unconceived Alternatives*, (Oxford: Oxford University Press).

Toulmin, S. (1981), 'Evolution, Adaptation, and Human Understanding', in Brewer, B. & Collins, B. (eds.) *Scientific Inquiry and the Social Sciences* (San Francisco: Jossey-Bass).

Tichý, P. (1974), "On Popper's definitions of verisimilitude", *The British Journal for the Philosophy of Science*, 25: 155-160.

Trigg, R. (1980), *Reality at Risk: A Defence of Realism in Philosophy and the Sciences* (Brighton: Harvester Press).

Turner, D. (2007), Making Prehistory: Historical Science and the Scientific Realism Debate (Cambridge: Cambridge University Press).

Urbach, P. (1983), 'Intimations of Similarity: The Shaky Basis of Verisimilitude', *The British Journal for the Philosophy of Science*, 34: 266-275.

van Fraassen, B. (1975), 'Wilfrid Sellars on Scientific Realism.', *Dialogue* 14 (1975), pp606-616.

van Fraassen, B. (1980), *The Scientific Image* (Oxford: Oxford University Press).

van Fraassen, B. (1983), 'Theory Confirmation: Tension and Conflict', in *Epistemology and Philosophy of Science: Proceedings of the Seventh International Wittgenstein Symposium*, pp. 319-329 (Vienna: Hoelder-Pichler-Tempsky).

van Fraassen, B. (1985), 'Empiricism in the Philosophy of Science', in Churchland and Hooker (1985).

van Fraassen, B. (1989), *Laws and Symmetry* (Oxford: Oxford University Press).

van Fraassen, B. (1994), 'Gideon Rosen on Constructive Empiricism', *Philosophical Studies* 74: 179-192.

van Fraassen, B. (1997), 'Structure and Perspective: Philosophical Perplexity and Paradox', in M. L. Dalla Chiara et al. (eds.) *Logic and Scientific Methods* (Dordrecht: Kluwer).

van Fraassen, B. (2001), 'Constructive Empiricism Now', *Philosophical Studies*, 106, pp151-170.

van Fraassen, B. (2002), *The Empirical Stance* (Yale University Press).

van Fraassen, B. (2006a), 'Structure: its Shadow and Substance', *The British Journal for the Philosophy of Science*, 57 (2006), pp275-307.

van Fraassen, B. (2006b), 'Replies to the Papers', in A. Berg-Hildebrand & C. Suhm (eds.) *The Fortunes of Empiricism* (Frankfurt: Ontos).

van Fraassen, B. (2007), 'From a View of Science to a New Empiricism', in Monton (2007) pp347-383.

van Fraassen, B. (2008), *Scientific Representation: Paradoxes of Perspective* (Oxford: Oxford University Press).

Votsis, I. (2004), *The Epistemological Status of Scientific Theories: An Investigation of the Structural Realist Account*, PhD thesis, London School of Economics, 2004, also at http://www.votsis.org/PDF/Votsis_Dissertation.pdf

Walton, D. (1991), *Begging the Question: Circular Reasoning as a Tactic of Argumentation* (London: Greenwood Press).

Weinberg, S. (1993), *Dreams of a Final Theory: The Scientist's Search for the Ultimate Laws of Nature* (Vintage Books).

Williamson, T. (2000), *Knowledge and its Limits* (Oxford: Oxford University Press).

Worrall, J. (1982), 'Scientific Realism and Scientific Change', *Philosophical Quarterly* 32, 201-231.

Worrall, J. (1984), 'An Unreal Image', *British Journal for the Philosophy of Science* 35, 65-80.

Worrall, J. (1985), 'Scientific Discovery and Theory Confirmation' in J. Pitt (ed.): *Change and Progress in Modern Science* (Dordrecht: Reidel).

Worrall, J. (1989), 'Structural Realism: The Best of Both Worlds?', *Dialectica* Vol. 43, no. 1-2 (1989).

Worrall, J. (1994), "How to Remain (Reasonably) Optimistic: Scientific Realism and the 'Luminiferous Ether.'" In D. Hull, M. Forbes, R. M. Burian (eds.), *PSA 1994*, vol. 1. (East Lansing, MI: Philosophy of Science Association).

Wray, K. Brad (2007), 'A selectionist explanation for the success and failures of science', *Erkenntnis* (2007) 67:81–89.

Zahar, E. (2001), *Poincaré's Philosophy: From Conventionalism to Phenomenology* (Chicago: Open Court).

www.ingramcontent.com/pod-product-compliance
Lightning Source LLC
Chambersburg PA
CBHW071328190426
43193CB00041B/947